T0140272

Studies in Computational Intelligence

Volume 725

About this Series

The series "Studies in Computational Intelligence" (SCI) publishes new developments and advances in the various areas of computational intelligence—quickly and with a high quality. The intent is to cover the theory, applications, and design methods of computational intelligence, as embedded in the fields of engineering, computer science, physics and life sciences, as well as the methodologies behind them. The series contains monographs, lecture notes and edited volumes in computational intelligence spanning the areas of neural networks, connectionist systems, genetic algorithms, evolutionary computation, artificial intelligence, cellular automata, self-organizing systems, soft computing, fuzzy systems, and hybrid intelligent systems. Of particular value to both the contributors and the readership are the short publication timeframe and the world-wide distribution, which enable both wide and rapid dissemination of research output.

More information about this series at http://www.springer.com/series/7092

Gul Muhammad Khan

Evolution of Artificial Neural Development

In Search of Learning Genes

Gul Muhammad Khan
Electrical Engineering Department
University of Engineering and Technology, Peshawar
Peshawar
Pakistan

ISSN 1860-949X ISSN 1860-9503 (electronic)
Studies in Computational Intelligence
ISBN 978-3-319-88435-6 ISBN 978-3-319-67466-7 (eBook)
https://doi.org/10.1007/978-3-319-67466-7

Softcover reprint of the hardcover 1st edition 2017

Printed on acid-free paper

This Springer imprint is published by Springer Nature
The registered company is Springer International Publishing AG
The registered company address is: Gewerbestrasse 11, 6330 Cham, Switzerland

Acknowledgements/Dedication

I dedicate this work to my Supervisor and Teacher in Life, Dr. Julian Francis Miller. The ideas and work presented here is the reflection of his guidance and his thinking of True AI. The whole concept is a culmination of our continuous discussions and his views of what Artificial Intelligence should be doing. Without him building some developmental concepts were almost next to impossible. He guided me during my Ph.D. and helped me develop systems capable of learning for itself, or at least the concept of it. Through his guidance I learned the very basics of life itself. I learned about myself, what am I? How I was built? Where I am in my body? How a single cell transformed into what I am today? The group behaviour of my cells in body and brain. I learned it all from you Julian, you are a true science to me.

Secondly I am thankful to Engr. Saba Gul and Engr. Rabia Arshad, My students and Colleagues who helped me review the book and transform it into a better shape.

And finally to all my friends and family members who supported and encouraged me to accomplish the task of completion of this book.

Contents

Declaration

The work presented in this book is the author's own work, unless it is stated otherwise. The following items have been previously published during this research.

1. Khan, G. M., & Miller, J. F. (2014). In search of intelligence: evolving a developmental neuron capable of learning. *Connection Science, 26*(4), 297–333.
2. Khan, G. M., Miller, J. F., & Halliday, D. M. (2011). Evolution of cartesian genetic programs for development of learning neural architecture. *Evolutionary Computation, 19*(3), 469–523.
3. Khan, G. M., & Miller, J. F. (2011). The cgp developmental network. *Cartesian Genetic Programming*, 255–291.
4. Miller, J. F., & Khan, G. M. (2011). Where is the Brain inside the Brain? *Memetic Computing, 3*(3), 217–228.

Hypothesis

Is it possible to implement an autonomous computational system inspired by neuroscience capable of continuously learning and adapting in complex environments.

Chapter 1
Making the Computer 'Brained'

For men at first had eyes but saw no purpose; they had ears but did not hear. Like the shapes of dreams they dragged through their long lives and handled all things in bewilderment and confusion. They did not knew of building houses with bricks to face the sun; they did not how to work with wood. They lived like swarming ants in holes within the ground, in sunless caves of the earth. For them there was no secure token by which to tell winter nor the flowering spring nor the summer with its crops; all their doings were indeed without intelligent calculations until I showed them the rising of stars and their setting, hard to observe. And further, I discovered to them numbering, pre-eminent among subtle devices and the combining of letters as a means of remembering all things, the Muses' mother skilled in craft. It was I who first yoked beasts for them, in the yokes and made of those beasts the slaves of trace chain and pack saddle that they might be man's substitute in the hardest tasks; and I harnessed to the carriage so that they loved the rein, horses the crowning pride of rich man's luxury. It was I and none other who discovered ships and sail driven wagons that the sea buffets. Such were the contrivances that I discovered for men.

—AESCHYLUS, Prometheus Bound

In his play Prometheus Bound; Aeschylus, a Greek dramatist describes how a titan, Prometheus freed mankind of ignorance and enlightened them with the gift of knowledge by deceiving the Greek god Zeus; and how this transgression incurred the wrath of Zeus. This brings in the light significance of knowledge and the fact that since time immemorial, there has been a struggle to acquire knowledge.

G.M. Khan, *Evolution of Artificial Neural Development*, Studies in Computational Intelligence 725, https://doi.org/10.1007/978-3-319-67466-7_1

1 AI-Defining Intelligence

We use the word "*Intelligence*" quite often but when asked to define the term, it puts us in a quandary. What is the basis for defining intelligence? What are the bases of factors that are used to define intelligence? Is it something that we are born with or does it come through our experiences and encounters in the physical world? But then are the sensations experienced by each individual similar? We are unable to quantify intelligent behavior as we are unable to define the basis for defining intelligence. We are unable to quantify intelligent behavior as we are unable to define the basis for defining intelligence. Intelligence is a relative term; an attempted definition of intelligence is; "To respond appropriately to a particular situation in a specific environment or under certain conditions." Artificial intelligence is the incorporation of intelligent behavior in machines.

2 Making the Computer 'Brained'

There is a famous saying that 'Necessity is the mother of invention', and the rise of science completely complements it. The rise of Science has also been due to the human necessities and their desire for fulfilling these necessities. With the developments in the field of Science and Technology, Human Desires and necessities have also increased proportionally. In the pre-historic days, human beings developed various tools for the sake of their safety and shaping their environment. Gradually human beings shifted their attention to controlling nature and environment. Science has provided us with the tools to do what used to be 'unthinkable' and 'impossible'. It has provided us with answers for the questions which were thought out to be great mysteries like weather conditions, planetary motions, economical development, and various other social, natural and cultural phenomena.

The major breakthrough in the field of Science and Technology was the development of Electronic Computers. It was that 'Magic Wand' which enabled us to reach the unprecedented heights. Now that Electronic Computer has provided us with various valuable tools for controlling and predicting our nature, scientists have turned their attention to developing computer programs which bear traits of human intelligence.

Recently, numerous Grand Challenges have been proposed in the USA and UK in the field of Computer Science. A Grand Challenge (GC) in computer research has already been established in UK.[1] As of now, there are eight Grand Challenges. GC5 "The Architecture of Brain and Mind" lays the foundation for various sub tasks which should be investigated for developing a robot which has the cognitive abilities of a human child. Sub task 1 deals with non-hierarchical description, design and framework of a series of computation models describing the functionality of the Brain.

[1] Brown, A., Furber, S., and Woods, R. "Grand Challenges in Microelectronic Design." A report supported by EPSRC Network grant, EP/DO54028/1. See http://intranet.cs.man.ac.uk/apt/uElecCV/ for further details.

This just deals with only a computational model. The aim of GC7 "Journeys in non-classical Computation" is the development of a mature science of all forms of calculations which can unite classical and non-classical issues. An integral component of this challenge is to refine, evaluate and understand the bio-inspired and complex systems. The Computing Research Association also held a conference in USA during the year 2003 which was called the "Grand Research Challenges" in Computer Science and Engineering. GC5 was concerned with conquering system complexity. The authors argued that such systems should be self-configuring, self-optimizing, self-maintaining, self-healing, self-protecting, self-differentiating and robust. According to the authors, achieving such systems will need an understanding and ability to control the emergent behaviors as well as the understanding of achieving learning in multi-agent systems. UK's funding agency, the EPSRC, has supported the research network which produced "Grand Challenges in Microelectronics Design".

Andrew Brown, Steve Furber, Rogers Woods along with their colleagues presented a report which listed and garnished these grand challenges. The last challenge that they offered was GC4 "Building Brains: Neurologically Inspired Electronic Systems." The GC4 can aid the GC5 through "developments in brain inspired novel computation." Systems which show intelligence, has the ability to learn, self-replication and self-adaptation are an inspiration to the idea of constructing a program which is modeled on brain. There are numerous features in human brain which are highly desirable but difficult to emulate in conventional computer systems. During the human life time, brain evolves and gradually gains the ability to deal with complex tasks encountered through development of abstract symbolic models. While keeping its integrity intact, the brain is adaptive and shows flexibility to any change in the environment by incorporating new experiences and responses to stimuli. Brain shows resilience to injuries by self-restoration and organizing mechanism.

The interconnected networks of similar neurons make the building blocks of brain which have the ability to learn and adapt. We believe that the seasoning of brain lies in the ability of neuron to develop and evolve, which highlights the importance of the research objective in acquiring a computational equivalent of neuron. Although inspired from the ability of the brain to evolve, the Artificial Neural Networks (ANNs) has not yet been able to reproduce many characteristics of biological neural systems (Gurney 1997) such as neural development, structure and mechanism of communication among neurons. The reason behind this was the lack of computation power for the model, however the modern growth in the computational power of computers along with our increased knowledge of neuroscience has made it possible to design more complex neuro-inspired approaches. ANNs deal with brains as a connectionist system such as a nodal network where every neuron serves as a node having the ability to process a signal. In biological neurons, before the arrival of signal at soma, complex signal processing takes place in neurite branches. Based on the signal received from the dendrites, decision about signal transformation is made. Since neurons are present in space, they are capable of redirecting signals to their neighbors by making a synapse through an electrochemical impulse (See Chap. 2, Sect. 5.4). The synaptic connections, the structure of branching and the number of neurons change with the passage of time resulting in the learning and adaptation ability of brain. The model discussed in this book has a dynamic structure which varies according to the task environment. The ultimate goal is the development of a system which has the ability to learn.

Biologically inspired computational systems have attracted the interest of researchers, where neural computation is not only considered from computing perspective but also from the perception of neuro science. Evidence clearly indicates that the sub-processes of neurons are dependent on time where different structures are rebuilt and altered. Memory is also not static and there is a constant gradual change in localization and mechanism for the stored (remember) information (Smythies 2002). The act of remembering is a process involving the reconstruction and the change in the primary structure associated with the particular event (Rose 2003). The physical topology of the neural structures keep on changing and plays an important role in learning ability of the brain (Kandel et al. 2000) (pp. 67–70). Koch and Segev (2000) have suggested that the "Dendritic Trees enhance the computational power". Dendrites are mostly responsible for shaping and integration of signals in complex manner (Stuart et al. 2001). They themselves cannot be simply credited for collecting and passing synaptic inputs to the soma (Stuart et al. 2001). The communication between Neurons take place through synapses which not only are the connection points, but they can also affect the shape and strength of the signal on short term (Kleim et al. 1998; Roberts and Bell 2002) as well as long term (Terje 2003). Over the years, there has been much proof provided about the importance of the physical structure of the Neuron. Biological neuron is very complicated on the basis of internal dynamics. That is why it is hard to replicate it into a dynamic computational model. There are also many processes in Biological Neuron that may not be necessary in the machine learning technique. However, to be able to identify essential subsystems for replicating the biological neuron in a computational model; this book assumes that the reader understands the gross morphology and connectivity of neurons (Alberts et al. 2002; Shepherd 1990). The conventional neural networks ignore the genetic makeup of neurons and its evolution process during learning; while it considers brain as a static connectionist system. Genetic Programming (GP) provide the concept of transferring genetic changes through the generations. Therefore Genetic Programming can be used for representing complex neuron engines which can be further advanced to behave like real neural systems going far beyond the boundaries of theoretical model of such systems. GP can solve such problems (Koza 1992). Such solutions show sudden incipient behaviours such as self-construction and self-reparability (Miller 2003, 2004) similar to living systems.

We explore Cartesian Genetic Programming (CGP) (Miller and Thomson 2000), a class of GP for developing the proposed computational network that is demonstrated to be capable of learning. An individual neuron is considered to be the computational unit comprised of chromosomes representing its sub-processing parts. The genotype of neuron is an assembly of chromosomes representing the integrant of neuron. The chromosomes are evolved to achieve the desired intelligent behaviour.

The model under discussion provides a neuron with structural morphology of dynamic synapses (Graham 2002), soma, dendrites (Panchev et al. 2002) and axons provided with branches. It employs the concept of synaptic communication for internal and external communication with neighboring branches. Branches attain a sense of virtual proximity by placing the neurons in a two dimensional toroidal grid. The branches can grow, shrink and communicate through the axon and dendritic branches. CGP chromosomes encode combinational digital circuits to idealize and represent the seven neural components (Khan et al. 2007). The chromosomes have the ability of encoding computational functions. This model allows the neurons, dendrites and axon branches to expand, perish and vary while attempting to unravel a problem. The information processing continue to be affected by synaptic morphology. Though this model is complex and there are a lot of parameters and variables involved, but it is getting closer to the complexity of the brain in search of intelligent behaviour. The seven chromosomes when run form a computational network, that can grow an assembly of neurons, neuritis and synapses having internal dynamics and environmental interactions of their own.

The learning ability of this network has already been tested on Wumpus World and checkers. The network exhibited learning proficiency in solving the wumpus world. According to (Hillis 1990) two agents in a co-evolutionary wumpus world can hardly perform tasks and they struggle with their own survival in a predator-prey relationship. There is a separate CGP computation network for controlling the agent. The "health" quality of agents may increase or decrease due to their decisions and experience in a two dimensional artificial environment. First agent can improve health by obtaining useful encounters and avoiding deleterious encounters. The other agent improves the health by head on confrontation with the first agent. According to (Paredis 1995), this is a "life time fitness evaluation" and discussed the manner in which this "arm race" can be a driving force for complexity.

Nolfi and Floreano also performed experiments which involved the seasoning of competing predators and prey robots to indicate that life time learning can help the individuals in obtaining ability for producing effective behaviour in varying conditions (Nolfi and Floreano 1998). The interesting part of this experiment is that both prey and predators keep changing from generation to generation, so both of them will face varying and more complex challenges. It was seen that in this situation, evolution can also display certain limitations. In spite of all of this, the research literature of evolved artificial developmental neural approaches is limited. These approaches will be reviewed and discussed in the section of neural development (see Chap. 4, Sect. 6). Our designed system can help the computational network make itself complex in response to its internal dynamics and environmental influences. The main reason behind this decision was the biological brain itself as it can do it without reshaping the underlying genetic makeup.

The basis of learning lies in the process of biological development. The development takes place with the passage of time. During the development, the system grows and the environmental interactions shape it. Biology proves that this emergent operation is initiated at the genetic level. Now the puzzling part is how the learning ability lies at the genetic level? This book has tried to entertain this question on the basis of two classical problems associated with artificial intelligence, i.e. Wumpus World and game of Checkers. A worthwhile object of AI research is to develop computer programs which can play games. Shannon (1950) presented the concept of utilizing a game tree which has a certain depth. He also suggested employing a board evaluation function which can allocate numeral score on the basis of better position of board.

According to (Dimand and Dimand 1996), Minimax is the method which can be used to determine the best moves. Samuel (1959) exploited it on computer checkers as a board evaluation function. As of now, the world champion in the game of checkers is Chinook (Schaeffer 1996) which employs "deep minimax search", "a huge database of end game positions" and "a handcrafted board evaluation function" based on human expertise.

Recently the Artificial Neural Networks (ANNs) have been employed for obtaining board evaluation functions for various games. The weights are adjusted on the basis of evolutionary techniques Othello (Moriarty and Miikulainen 1995), Go (Richards et al. 1998), Chess (Kendall and Whitwell 2001), and Checkers (Chellapilla and Fogel 2001).

From literature of computer games, many effectual methods (e.g. minimax, board evaluation function) are introduced. Human beings on the other hand don't use such methods, instead they consider only few probable moves, and the decision making regarding 'how to play' is highly instinctive and empirical. The computer programs also learn the game just like a human being learns to be better at a game. But how is this possible? The main interest of our work is to ascertain the rise in learning ability and how it can be incorporated in a genotype, which after execution results in a neural network with the ability to play a game at a higher standard of decision making. Every player contains a genotype which develops into a computational neural framework during the game. Our method uses only a few of the traditional perceptions which are utilized in the Artificial Neural Networks. Most of the aspects of the neural functions are primarily due to the genotype's evolution.

To test the ability of our approach for learning, the model of this book made two seasoned players compete against each other utilizing a form of co-evolution. This model also tested the exhilaration each player for competing against a minimax based checker program (MCP). During the previous case, the two agents were loaded with Cartesian Genetic Programming Developmental Network (CGPDN) and then they were evolved together for a number of generations. After performing many evolutionary runs of the system, all of the more evolved agents were tested against the less evolved ones. With the result being in favour of the more evolved agents. In later case, when the player is evolved in competition with a MCP, the more evolved players were once again tested with the lesser evolved players with more evolved players out performing them. This clearly indicates that the evolution with the passage of time produces programs which grow into a better checker playing systems.

Further analysis of self-configuration and experiential learning abilities of the system exhibited interesting results. The system had the tendency to optimize its neural architecture, and manipulate random number of inputs and outputs at execution.

In Chap. 2, a review of the essential biology and neuroscience which lay the foundation for the model is discussed. Chapter 2 also provides a detailed background of the biological brain, neurons, different Compartments of neurons, neural signal processing and learning in brain.

Chapter 3 provides a review of computational evolution and gives a deep insight into genetic programming technique called Cartesian Genetic Programming which has been used in our work.

Chapter 4 highlights all the relevant work done in the sphere of Artificial Neural Networks (ANNs). It also discusses the various types of ANNs, neuro-evolution, neural development techniques, and catastrophic forgetting.

Chapter 5 will give a comprehensive description of the model explored, explaining the layout of the neurons, their internal architecture, their interface with the environment and internal signal processing.

Chapter 6 will highlight various methods of application of the developmental model presented in this research, to the 'Wumpus World problem' for demonstration of its learning abilities. Both evolutionary and co-evolutionary methods are explored to obtain genes that can create learning abilities in neurons and ultimately the systems they are installed in.

Chapter 7 will give an insight into the methods of playing checkers using the model of this work. The network starts with simple random genes and random neural architecture. Later on it develops and learns the method of playing checkers with the evolved genes that develops a neural architecture capable of learning. The training of system is carried out through co-evolution and against a minimax based checker software program.

Chapter 8 will present the conclusion and discussions on the the present work and highlight the plans for future work.

Chapter 2
The Biology of Brain: An Insight into the Human Brain

Human brain is a very complex but fascinating subject. The structure of brain has always been an attractive topic to the human beings. This small mass miraculously controls every action of the human body. Unfortunately it is not possible to cover every single aspect of the human brain in just one book. This book will try to explain various aspects of brain which are thought to be the main cause of the learning ability of the brain. Brain itself is made up of small building blocks called the "neurons" that are a major source of attraction on their own. The complexity of neurons, its structure and its functionality is a debated topic. Neurons are spread throughout the body and are present in different shapes, however their basic mechanism does not change (Kuffler et al. 1984). Figure 1 shows various types of neurons that exist in the human nervous system. This is why; we are going to begin with a detailed insight into The Human Nervous system.

1 Human Nervous System

The Human Nervous System is comprised of two main parts:

Central Nervous System: The information from sensory neurons is carried to the Central Nervous System (CNS), where it is processed and sent to the desired motor neurons. The motor neurons are responsible for controlling various physical activities of humans. The CNS is responsible for these functionalities as well as creating a memory of these processes for future reference.

Peripheral Nervous System: It is made up of the motor and sensory neurons. Once the sensory neurons send the information about the environment to the CNS, the CNS then processes the signals. The desired behaviour is visible in the form of various human activities controlled by motor nerves through signals. The output of Motor Neuron is visible in the form of the human body's motor action.

G.M. Khan, *Evolution of Artificial Neural Development*, Studies in Computational Intelligence 725, https://doi.org/10.1007/978-3-319-67466-7_2

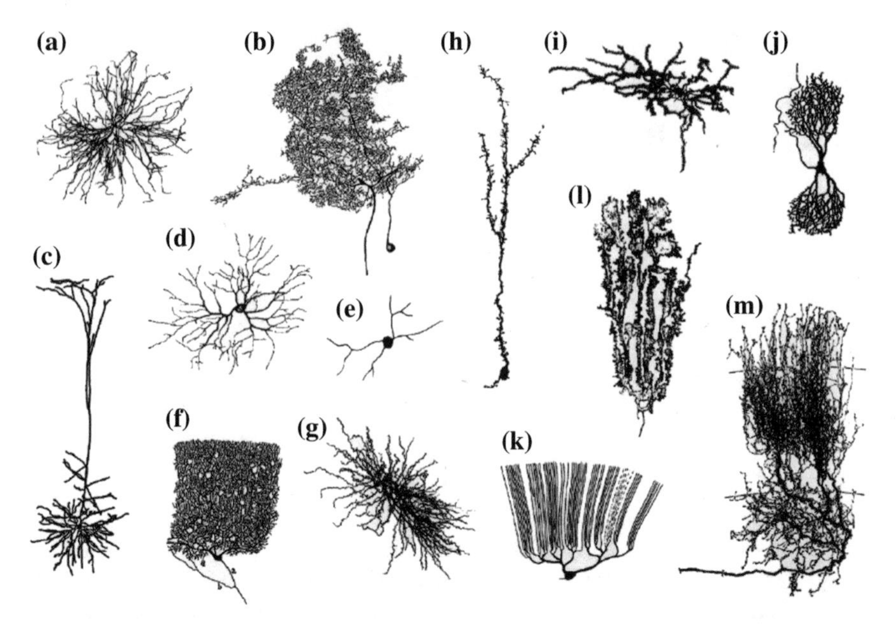

Fig. 1 Different types of Neurons, taken from (Mel 1994), **a** Alpha motorneuron in spinal cord of cat, **b** Spiking interneuron in mesothoracic ganglion of locust, **c** Layer 5 neocortical pyramidal cell in rat, **d** Retinal ganglion cell in postnatal cat, **e** Amacrine cell in retina of larval tiger salamander, **f** Cerebellar Purkinje cell in human, **g** Relay neuron in rat ventrobasal thalamus, **h** Granule cell from olfactory bulb of mouse, **i** Spiny projection neuron in rat striatum, **j** Nerve cell in the Nucleus of Burdach in human fetus, **k**. Purkinje cell in mormyrid fish, **l** Golgi epithelial (glial) cell in cerebellum of normal-reeler mutant mouse chimera, **m** Axonal arborization of isthmotectal neurons in turtle

2 Central Nervous System (CNS)

The Central Nervous System is responsible for controlling all the processes of the human body and producing actions in response to the environmental inputs. These inputs can be in different forms such as pressure on muscle receptors, visual input of eyes, chemical inputs through taste, auditory and/or olfactory inputs. There are more than 10^{11} neurons in the central nervous system. These neurons form tens of thousands of connections with other neurons.

The outer area of the human brain is a grey layer, which is only a few millimetre in thickness. This grey layer contains neuron bodies. This area is termed as cerebral cortex or simply cortex (Kandel et al. 2000). The cortex has a large surface area, and is highly folded, and as described by (Kandel et al. 2000) the most complicated, most studied and fascinating part of the brain. The next layer of brain is mostly made up of myelination and is termed as the white matter. The insulation to the axons of neurons is provided by this layer. It also supports the structure. Figure 2 depicts different regions of the brain which are explained below.

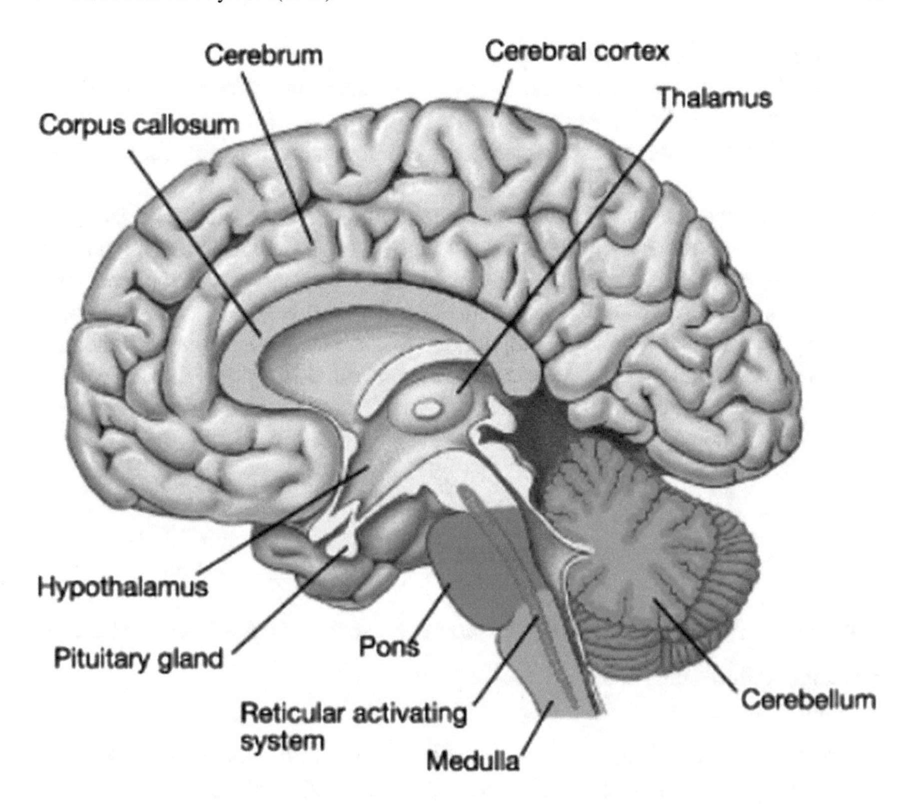

Fig. 2 Schematic diagram of various regions of the brain, taken from http://wps.prenhall.com/wps/media/objects/1928/1974895/f04-10.gif

- Heart rate, breathing, peristalsis, and various reflexes such as swallowing, coughing, sneezing and vomiting are controlled by the Medulla.
- The temperature homeostasis, water homeostasis and the release of hormones of pituitary gland is controlled by the Hypothalamus.
- Different hormones are secreted by the Pituitary Gland also known as the Master Gland.
- Cerebellum is responsible for the coordination of muscle movement; hence controlling balance, posture, walking, jumping and running.

2.1 The Cerebral Cortex

Different functions of cortex are localized into discrete areas, which can be divided into three groups (Kandel et al. 2000):

- Sensory Areas: The sensory inputs from the sensory organs are received and processed by the sensory areas. Every sense organ has different sensory areas such as visual, auditory, smell and skin.
- Motor Areas: Those areas which organize and send motor outputs to skeletal muscles.
- Associative Areas: These areas are involved in higher processing. The short term memories called sensory maps are produced by the associative areas for real world experiences. There are multiple copies of sensory maps present in the associative areas that change as the sensory map changes. These copies are of great importance as they can be used to either compare or associate sensory inputs with the prior experiences and then make decisions based on them. So they play their role in various advanced skills namely visual recognition, language understanding, speech, writing and memory retrieval.

2.2 *Types of Cells in the Human Brain*

According to (Kandel et al. 2000), the Nervous System consists of two types of cells:

- Ganglia (Glia) Cells, and
- Neurons

There are about 100 billion neurons in the brain, while glial cells are 10 to 50 times more than the number of neurons in the brain. Glial cells are responsible for maintaining the structure of the nervous system. They also provide assistance in maintaining a lossless flow of electrical spikes in the axon through the provision of electrical insulation to the axon. Glia cells also constitute the myelin sheath around the axon. The processing of information in the brain is done by the neurons. According to (Kandel et al. 2000) neurons are different from the rest of the cells in the body as they differ both in functionality as well as the biophysical structure.

The next section will describe the neurons in detail.

3 Neurons

Neurons are the building blocks of brain. According to (Kuffler et al. 1984) the appearance and structure of neurons vary based on their location in the brain (as shown in Fig. 1). However, the basic structure of neuron is same. There are three main parts in the neuron (as shown in Fig. 3).

- Dendrites: They are the inputs of the neuron. They receive the information sent by other neurons and then transfer it to the cell body. The structure of dendrites is just like a tree. Their branches are close to the cell body.

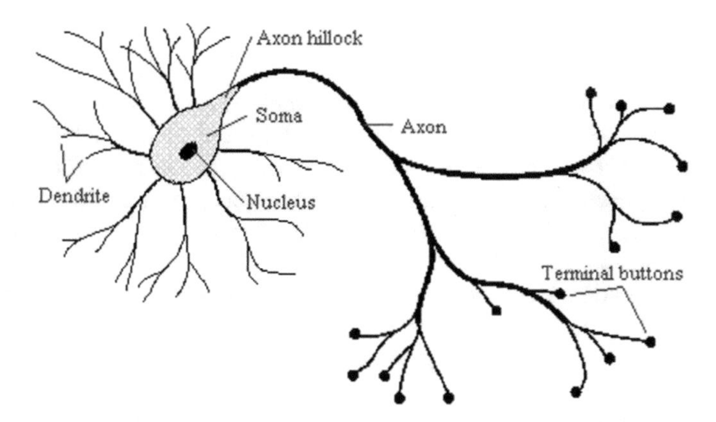

Fig. 3 Schematic of neuron showing its different parts: dendrite, soma, axon, axon hillock and axon butons. Taken from http://www.virtualventures.ca/~neil/neural/neuron1.gif

- **Axons:** Axons are considered to be the outputs of neurons. The transfer of information from one neuron to another takes place through the axons by propagation of spike or action potential. Axons make branches away from the cell body. They make synapses or connections with the dendrites and cell bodies of other neurons.
- **Cell body:** It is the area which performs the main processing for the neurons. The instruction from dendrite branches is received in the form of electrical disturbances which are then converted into action potentials. They are then transferred to other neurons through the axons. The development of neurons and branches is also controlled by the cell body.

Compartments of neurons are further elaborated in the following subsections:

3.1 Dendrites

Dendrite comes from the Greek word "dendron" which means tree. As described by the name, the structure of dendrite resembles that of a tree (as shown in Fig. 3). Dendrites have spines that make synapses with the axon of other neurons. Synapses are responsible for the reception of the bulk of electrical signals from other neurons. Dendrite spines can also be used for reception of signals from some dendrites. There are channels present along the length of dendrites which are responsible for the modulation of signals either through amplifying (voltage gated channels) or attenuations (leaky channels) (Alberts et al. 2002). The membrane channels are biological membrane proteins. This allows the movement of ions, water and other solutes for passively passing through the membrane due to an electrochemical gradient.

3.1.1 Dendrite Spine

It is a small (sub-micrometer), thrust out membrane. It extends out from a dendrite and half of a synapse is made up of dendrite spine (Nimchinsky et al. 2002). The spine contains a number of receptor channels. The opening and closing of the channels is based on signals received from the axon buttons (as shown in the Fig. 3) of the other neuron through the neurotransmitters. Neuro-transmitters are chemicals and are responsible for communication between the neurons.

Neuro-transmitters act on the receptor channels at the dendrite spine, which causes the channels to open. After the channels are opened, the ions flow in and out of the spine due to which a disturbance is produced in the electrical potential of the dendrite spine.

Spines are present in the dendrites of pyramidal neurons, the medium spiny neurons of the striatum and purkinje cells. They are responsible for the modulation of the signal which comes from other neurons. Spines have a variety of shapes and these shapes are in accordance with the distinct evolutionary stages and strengths of a synapse (Nimchinsky et al. 2002).

3.2 Axon

The long filament which extends from the cell body (the soma) of the neurons is called the axon. It is responsible for carrying nerve impulses away from the soma to the presynaptic terminal buttons. These impulses are then transmitted to other neurons. However if we are dealing with motor neurons, then the impulses will be transmitted to the muscles. Axons are usually long and can be up to 1 m in length. The impulses are carried at the speeds of 100 m/s or more. According to (Michael et al. 1998), the transmission speed of the nerve impulse is directly proportional to the axon's diameter. Many axons can be seen through the naked eye.

The insulation is provided to most of the neurons through myelination (as shown in Fig. 4). The axons are folded in myelin which assists in flow of signal in the axon.

The myelin sheaths are present at discrete points around the axons, thus some breaks are left behind (as shown in Fig. 4). These breaks are called nodes of Ranvier. The active transmission of signal along the axon is made possible through these nodes. The membrane channels are exposed at these nodes, which results in the jump of signal from node to node. This results in an increased conductivity velocity. The Ranvier nodes contain sodium channels and can serve in preventing the decay of nerve impulses by amplifying them (Kandel et al. 2000). The amplification is done through the action potential firing in the nearby node. The jumping of action potential from one node to another is called "saltatory conduction" which was suggested by Ralph Lillie in 1925 (Lillie 1925). The first empirical affirmation of saltatory conduction was provided by ichiji Tasak and Taiji Takeuchi, Alan Hodgkin and Robert Stmpfli (Tasaki 1939; Tasaki and Takeuchi 1942; Huxley and Stmpfli 1949).

There are two main parts of an Axon

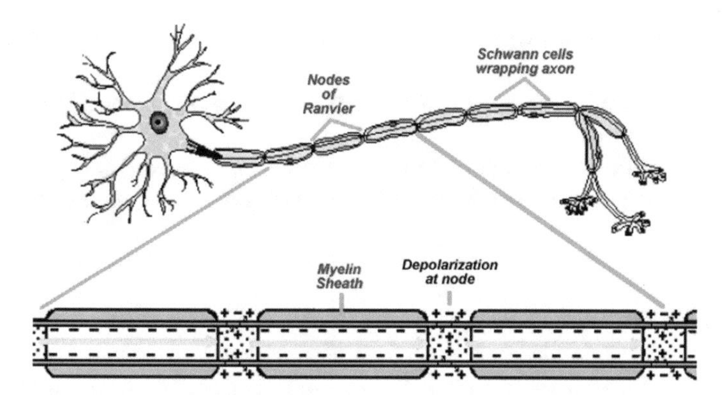

Fig. 4 Schematic of neuron showing nodes of Ranvier and myelination. Taken from http://www.coolschool.ca/lor/BI12/unit12/U12L03/Saltatorycondn.png

- Presynaptic Terminals
- Axon Hillock

3.2.1 Presynaptic Terminals

Axons also have projections called buttons. They perform the task of a presynaptic terminal (as shown in Fig. 3). They are present either at the end of an axon or along the length of axons (Kandel et al. 2000). The classification of axonal buttons can be performed based on their synaptic vesicle characteristics. An individual button can form single or multiple synapses with their postsynaptic partners.

3.2.2 Axon Hillock

They are present at the axon base, and the number of ion channels present there is greater. As a result of which, the probability of occurrence of firing is most in this region (Kandel et al. 2000).

3.3 Summary of Differences Between Axons and Dendrites

There are many differences between axons and dendrites. Some of them are listed below:

- The job of the dendrite is to transfer information to the cell body while the axons transfer information away from it.
- The surface of axon is smooth while the surface of dendrites is rough.

- Dendrites are not insulated by the myelination while axons have myelin sheet around it.
- The branching of dendrites is near the cell body while the branching of axons is distant from the cell body.
- Dendrites work as receivers due to the neuro-receptor release sites while axons work as transmitters due to the presence of neuro-transmitter release sites.

3.4 Synapse

Synapse is the junction between the two neurons (Shepherd 1990). Due to the movement of ions across the gap, the presynaptic electrical signals are transferred to a post synaptic terminal at the synapse. The nerve impulse received from the soma opens the voltage gated channels at the pre-synaptic terminal. This results in the inward flow of the calcium ions. Figure 5 shows voltage gated calcium channels.

The presence of calcium ions is responsible for the fusion of the synaptic vesicles with the cell membrane. Then the contents i.e., the neurotransmitter chemicals are released through "exocytosis" (a cellular process through which the cells excrete by products or chemical transmitters) as shown in Fig. 5. The neurotransmitters then diffuse across the synaptic cleft where they are bound for the neuro-receptors in the post synaptic membrane. This results in the opening of channels causing the inward flow of sodium ions in the post synaptic terminal. Depolarization of the post synaptic cell membrane takes place. In this way the information is transferred across synapse from one neuron to other.

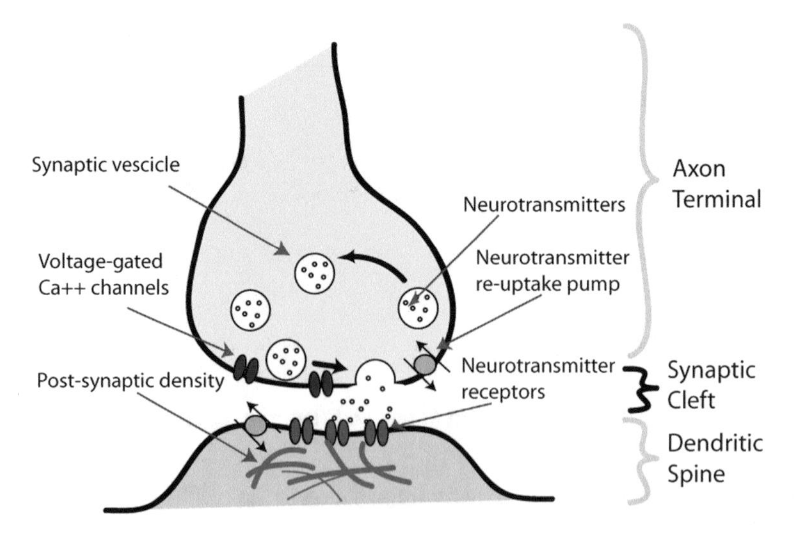

Fig. 5 Schematic of a synapse showing pre and post synaptic terminals and synaptic cleft and internal dynamics. Taken from http://upload.wikimedia.org/wikipedia/en/4/46/SynapseIllustration.png

3.4.1 Different Parts Synapse

There are three constituents in a synapse (Kandel et al. 2000).

- Pre-synaptic Terminal
- Synaptic Cleft
- Post Synaptic Terminal

They are further detailed below:

Presynaptic Terminal

The synaptic vesicles are present in the presynaptic element. The synaptic vesicles are filled with neuro-transmitters (as shown in Fig. 5). There can be variation in the size, shape and content of the vesicles on the basis of type of synapse.

Synaptic cleft

It is a 20–30nm wide separation between the presynaptic and postsynaptic membranes. It is usually filled with a dense plaque of inter-cellular material (as shown in Fig. 5).

Post Synaptic Terminal

It is mostly the dendrite spines which consist of the neuro-receptors (as shown in Fig. 5). The opening of channels and the influx of ions to the dendrite spines is due to the combination of the neuro-receptors and neuro-transmitters. Every neuro-receptor reacts to a particular neuro-transmitter.

4 Electrical Signaling

4.1 Membrane Biophysics

Movement of ions is responsible for the processing of information in biological organisms. These ions are either cations like Sodium (Na^+), Potassium (K^+), Calcium (Ca^{2+}) or anions like Chloride (Cl^-) (Mummert and Gradmann 1991). The flow of ions can take place either through diffusion or electric fields. The difference in the concentration of ions between the two regions, results in the diffusion of ions. The difference in the concentration of electric fields between the two regions can also result in the flow of ions. The membrane around the cell is impermeable to ions,which makes it difficult for the ions to diffuse through the membrane. This causes a net potential difference across the membrane (Lieb and Stein 1986). The ions can move across the membrane through two different mechanisms which are discussed below.

4.1.1 Membrane Channels

The membrane channels are biological membrane proteins which permit the passive movement of ions, water and other solutes to pass through the membrane down towards their electrochemical gradient (Alberts et al. 2002). Channels can be in two different states. They will either be open to let the flow of ions take place through diffusion or they can be fully closed. The fully closed channels can be voltage-sensitive channels which open when a certain threshold voltage is established across the membrane. It can also be Ligand-Gated Channel which opens in response to a ligand neuro-transmitter. Ligand is a specific chemical. There are also certain channels which are sensitive to light, temperature and pressure.

4.1.2 Ion Pumps

Every animal's cell membrane contains a protein pump known as the Na+K+ATPase pump (Kandel et al. 2000). This pump is responsible for continuous pumping out of 3 sodium ions (Na^+) from the cell and pumping in 2 potassium (K^+) ions (as shown in Fig. 6). If this process keeps on going, then there will be no potassium or sodium ions left for pumping. However, there exist some leaky channels as well (see Sect. 2.6.2). The concentration of different ions vary inside and outside the neurons such as the concentration of Potassium ions outside the neuron is 20 times greater than the concentration of Potassium ions inside, and the concentration of sodium ions inside is 9 times greater than the concentration of sodium ions outside the neuron (Steinbach and Spiegelman 1943).

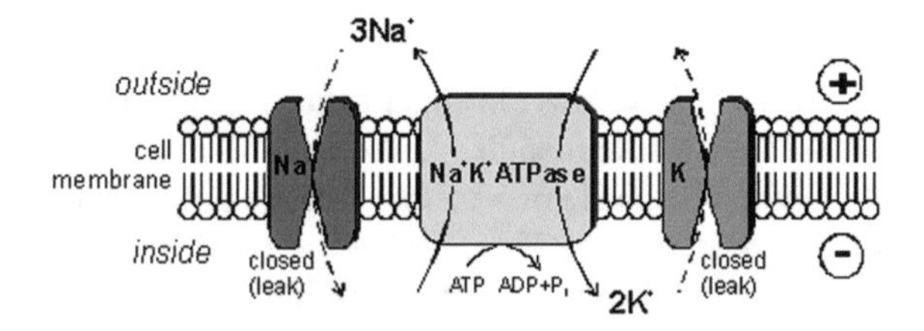

Fig. 6 Schematic of ATP pump

4.1.3 Membrane Potential

Na+K+ATPase pump along with the leakage channels are responsible for causing a stable imbalance of Na^+ and K^+ ions across the membrane. This imbalance results in a potential difference across the cell membranes of all animals. This potential difference is called the membrane potential which is always negative inside the cell and its magnitude varies from −20 mV to −200 mV in different cells and animal species.

4.2 *Resting Ion Channel*

There are resting ion channels present in the biological neurons which are open all the time (Kandel et al. 2000) (page 125–139). These are also known as the leakage channels or leak channels. There is an electrochemical gradient present in the resting ion channels which allows the flow of ions across the membrane. The positively charged Potassium ions (K^+) are allowed to flow outwards while the negatively charged Chloride (Cl^-) ions move into the cell through the electrochemical gradient. The resting membrane potential is hyperpolarized or negative due to the net efflux of positively charged ions as the number of resting sodium channels is less compared to the resting potassium and chloride ions. This continuous flow of ions is termed as the "leakage current" which is much smaller in magnitude as compared to the currents flowing through the voltage-gated ion channels. The flow of ions, continues across these channels till the net current becomes zero which indicates that the equilibrium potential is attained. The equilibrium voltage can be represented by the Nernst Equation (Bernstein 1902).

$$E = \frac{RT}{nF} \ln \frac{[\text{outside ion concentration}]}{[\text{inside ion concentration}]}.$$

where;

n = The valence charge of the ion such as +1 for K^+, +2 for Ca^{2+} and −1 for Cl^-)

T = The temperature in Kelvin

R = Molar gas constant

F = Total Charge of a mole of electrons

According to the Nerst equation, the equilibrium voltage of Potassium ((E_K)) is −75 mV while the equilibrium voltage of Sodium (E_{Na}) is +55 mV. These potentials show that these currents can never be zero, however at some voltage E_m the net current of all ions across the membranes can be zero as indicated by the Goldman Equation (Goldman 1943).

$$E_m = \frac{RT}{F} \ln \left(\frac{P_{\text{K}}[\text{K}^+]_{\text{out}} + P_{\text{Na}}[\text{Na}^+]_{\text{out}} + P_{\text{Cl}}[\text{Cl}^-]_{\text{in}}}{P_{\text{K}}[\text{K}^+]_{\text{in}} + P_{\text{Na}}[\text{Na}^+]_{\text{in}} + P_{\text{Cl}}[\text{Cl}^-]_{\text{out}}} \right)$$

where;

P_{ion} = The permeability

$[ion]_{out}$ = The extracellular concentration

$[ion]_{in}$ = The intracellular concentration of the ion

The equilibrium potential E_m also known as resting potential is typically −70 mV.

4.3 *The Action Potential*

Neurons can produce the 'action potentials' whose nature was discovered by Alan Lloyd Hodgkin and Andrew Huxley in 1952 (Hodgkin and Huxley 1952). It is also called a nerve impulse and is formed due to signals from other neurons. Other neurons produce these signals at their dendrite spine either through electro-chemical synapses or pressure or chemicals. The production of action potential takes place due to opening and closing of voltage sensitive channels as a result of changes in membrane voltage V_m. The change in membrane voltage also results in the change of membrane's permeability. The Goldman equation clearly indicates that the variation in ionic permeability can cause a change in the equilibrium potential E_m and the membrane voltage V_m (Goldman 1943). There is a positive feedback which arises from interaction between Vm and ion channels. This positive feedback is responsible for rising phase of the action potential. Most of the voltage-sensitive sodium ion channels are opened by raising V_m. An action potential has following four main phases (as shown in Fig. 7).

- Rising phase
- Falling phase
- Undershoot phase
- Refractory period

4.3.1 Rising Phase

Action potentials rise from the influx of sodium cations into the cell which increases V_m. This results in opening of more voltage gated sodium channels to open which increases membrane permeability to sodium ions. As long as the sodium channels are being charged and V_m does not approach E_{Na} (55 mV), the positive feedback continues as shown in Fig. 7. This process usually takes place at the axon hillock.

4.3.2 Falling Phase

When the membrane potential gets closer to E_{Na}, it causes a reduction in the activeness of the sodium channels; resulting in the reduction of the membrane's perme-

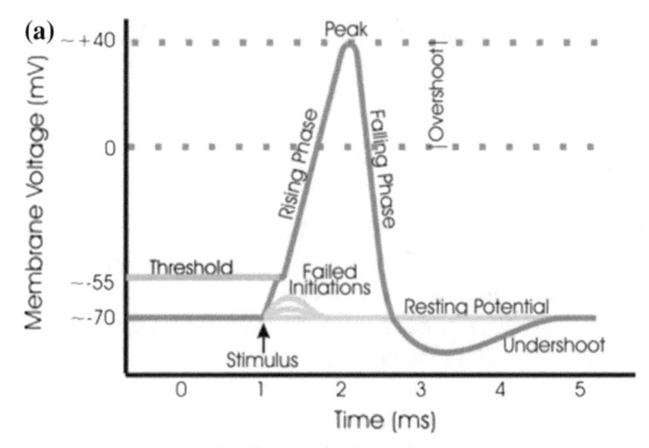

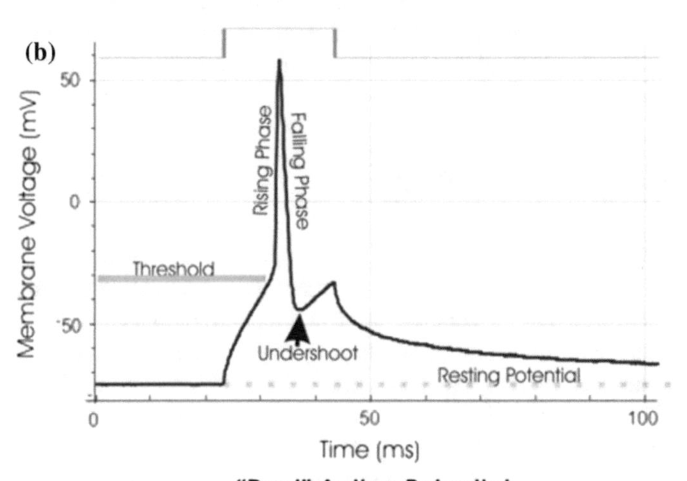

Fig. 7 Schematic and real action potential curve, taken from http://upload.wikimedia.org/wikipedia/commons/thumb/c/cc/Action-potential-vert.png/422px-Action-potential-vert.png

ability to sodium. It also causes the movement of V_m towards resting potential. This marks the end of the falling phase. This is the point where the potassium channels start to open a bit more, which causes the permeability of potassium to increase. It also drives V_m towards E_K. This phase is also called repolarization.

4.3.3 Undershoot Phase

During the falling phase, some of the potassium channels remain open even after the membrane potential reaches its normal resting potential. This causes the membrane voltage to go below the resting potential as shown in Fig. 7. This process is termed as undershoot or hyperpolarization.

4.3.4 Refractory Period

When the membrane potential reaches its resting potential, then most of the channels reach the refractory state. Since they have not returned to their normal state, therefore it is cumbersome to fire a new action potential. To be able to quickly recover to the firing state, the membrane potential has to stay in hyperpolarized state for a longer period of time. This time during which the neuron cannot fire is known as the refractory period, and the period in which there is no new action potential; that is called the "absolute refractory period". There is also a period in which it is more troublesome to produce action potential and is termed as "relative refractory period". Due to these refractory periods, the transfer of action potential is always in a single direction, i.e., it moves away from the soma, along the axon. A stronger stimulus will make the neuron fire quickly by overcoming the relative refractory period.

4.4 Sub Threshold Behavior

The signals are usually initiated at the dendrite spines where the cations are injected because of the chemical synapses, sensory neurons or pacemaker potentials. There are two phases termed as the depolarization and repolarization. Depolarization takes place when the neurotransmitters, released by the neurons, move across a synaptic cleft and bound themselves to specific proteins (neuro receptors) on the post synaptic terminal of the neuron. This results in a variation in the ion channels of the post synaptic terminals, causing the opening of channel and the flow of ions through the pore. These ionic currents can be caused by Sodium (Na^+) and Calcium (Ca^{2+}) ions which results in depolarization. The ionic currents can also result from the influx of Chloride (Cl^-) ions or outward flow of Potassium (K^+) ions causing hyper polarization. Depolarization takes place when the normal voltage polarity (negative inside) becomes positive while hyperpolarization takes place when the voltage polarity becomes more negative. Hyperpolarization takes place due to inhibitory synapses. This restores the original polarity and is known as repolarization. This signal is then transferred to soma through the dendrites. If the net effect (potential) is above a certain threshold (soma threshold), then an action potential will be produced. However, if it is below the threshold value; the decay of the signal will start until it is incremented by a stimulus coming from another dendrite.

4.5 Cable Theory

This theory was originally developed by Lord Kelvin for transatlantic telegraph cables (Kelvin 1855). Then Hodgkin and Rushton used it for modeling of the signaling in neuron (Rall 1989). This model considers a neuron as cylindrical transmission cable which has passive properties. It can be represented through the differential equation:

$$\tau \frac{\partial V}{\partial t} = \lambda^2 \frac{\partial^2 V}{\partial x^2} - V$$

Where
V = Voltage across the membrane.
t = Time
x = The position along the length of neuron
λ = The characteristic length and τ = The characteristic time at which the voltage decays in response to a stimulus.

Figure 8 presents the circuit diagram of the model. It is the compartmental model of a cable. The characteristic length and time scale can be determined in terms of resistance and capacitance per unit length using the following relations:

$$\tau = R_m C_m$$

$$\lambda = \sqrt{\frac{R_m}{R_l}}$$

where;
R_m = The membrane resistance
C_m = The capacitance due to electro static forces
R_l = The longitudinal Resistance.

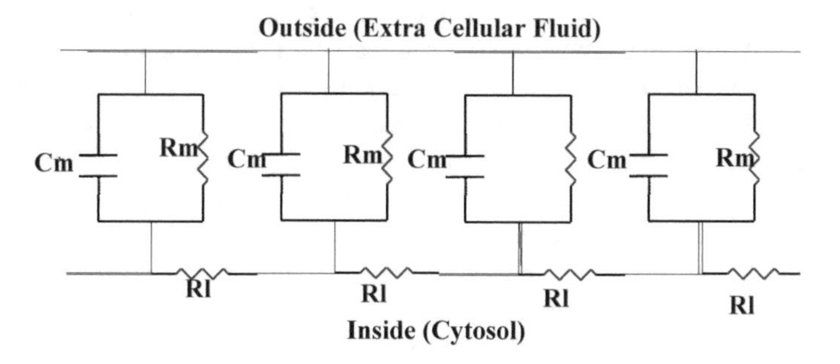

Fig. 8 Cable theory's simplified view of a neuronal fiber. The connected RC circuits correspond to adjacent segments of a passive neurite. The extracellular resistances R_e (the counterparts of the intracellular resistances R_i) are not shown, since they are usually negligibly small; the extracellular medium may be assumed to have the same voltage everywhere, taken from http://en.wikipedia.org/wiki/Action-potential-Cell-membrane

The effect of the diameter of the neuron on conduction velocity in fibers which have no myelination can be seen from the above mentioned parameters. As a consequence, it can be seen that the resistance is inversely proportional to the diameter and there is a direct proportionality between the capacitance and diameter of a neurite. In this section we discussed the signaling mechanism within and between the neurons. The next section will discuss the learning and memory capabilities in the brain from the context of signal modulation and developmental process.

5 Learning in the Brain

Neural processes are function of time. Neuron has a highly complex architecture inside (Smythies 2002). The synapses between the neurons can be replaced, deleted and there can even be a creation of new synapses. These variations occur more often compared to the anatomical changes in the brain. The variations take place actively, where new connections are made and the unused brain connections are discarded. The locations and mechanism for the remembered or stored information are in constant though largely gradual change (Rose 2003). There are some variations in a specific area of brain which takes place due to the learning experience. This persists for a couple of hours. It is because of the shifting of the number and the position of a few dendrite spines on a few neurons in a specific region of brain. The physical structure of neuron topology inside the brain constantly reshapes and is a fundamental part of acquiring the ability to learn (Kandel et al. 2000). The capability of brain to develop hierarchical structure gives it the ability to react to changes in the environment along with adapting to new information and growing in relation to the variation in the physical development of the individual (Hawkins 2004). The ability of brain to produce these memory structures is dependent upon the internal processes involved at the neuron level. To date; habituation, sensitization and associative learning have been identified as the types of learning (Wood 1988). All of these have shown the tendency to be for short term, ranging from the order of minutes to hours. They have also shown long term characteristics which have the order of days or weeks, but the mechanism for each of them is different. Short term memory is also accounted for by changes in ion concentrations and ion pathways in the pre and post synaptic sides of the synapse. On the other hand, the anatomical changes in the synaptic connections give rise to the long term learning. The model discussed in this book comprehends both short term and long term memory development mechanisms. For the modulation of signals and neural development for long term memories, a weighted adjustment mechanism has been used. The next subsection provides a detailed description of various kinds of learning mechanisms in brain.

5.1 Synaptic Plasticity

Synaptic Plasticity means the variability of the strength of a signal transmitted through a synapse (Debanne et al. 2003). It might also involve the variation of the post-synaptic neuron's excitability (the chance that it will fire after the given level of stimulus) in response to the amount of stimulus it might have received in the past (Gaiarsa et al. 2002). The receptors on the membrane can also be varied by synaptic activity (Frey and Morris 1997). Experiments have indicated that a permanent developmental change to neural architecture can also take place due to the development of a positive feedback loop. The positive feedback loop is developed due to the reinforcement of strength of synapse by stimulation or weakening of the strength of synapse due to the lack of stimulation. The positive feedback causes some of the cells to never fire while others firing a lot (Gaiarsa et al. 2002).

5.2 Hebbian Theory

This theory is about the basic mechanism of synaptic plasticity. The synaptic variation occurs from the presynaptic cell's recurring and sustained stimulation of the postsynaptic cell. This theory was proposed by Donald Hebb in 1949 (Hebb 1949). According to this theory: "When an axon of cell A is near enough to excite a cell B and repeatedly or persistently takes part in firing it, some growth process or metabolic change takes place in one or both cells such that A's efficiency, as one of the cells firing B, is increased." The theory is also stated as "Cells that fire together, wire together" (Shatz 1994). This form of learning is also termed as activity dependent synaptic plasticity. The work of Donald laid the foundations for Synaptic Plasticity in Spiking Neural Network (SNN) models (Chap. 4 Sect. 4.2.3); the basic doctrine of this model is "the repeated and persistent stimulation of a postsynaptic neuron by a pre-synaptic neuron" (Hebb 1949). This results in strengthening of the connection between two cells.

Hebbian learning is also of use in the "Spike-timing-dependent plasticity (STDP)"' (Roberts and Bell 2002). STDP relies on the relative timing of the pre-synaptic and post-synaptic action potentials. This type of synaptic modification automatically results in balanced synaptic strengths for making the postsynaptic firing irregular. However it makes it more sensitive to presynaptic spike timing. The synapses modifiable by STDP compete for control of the timing of the postsynaptic action potentials. Any stimulus which can cause post synaptic neuron to fire within short intervals of time is also able to develop strong synapses.

STDP rules can be utilized to update the weights in SNN networks (Roberts and Bell 2002). The changes in weight are dependent upon the relative timing of the pre- synaptic and post-synaptic spikes. Different rules were proposed by Song for modifications in the STDP (Song et al. 2000). The modification of synapses takes place whenever there is correlation between presynaptic and postsynaptic activity according to STDP. However such correlated activity occurs only once in a while and it happens by chance. That is why ANN's must utilize the covariance based modification rather than opting for correlation based modification. According to Song, the synaptic weakening through STDP is more efficient than the synaptic strengthening. This indicates that the uncorrelated synapses are weakened by the STDP and the synapses which can provoke the action potential in the postsynaptic neurons are strengthened. These systems rely on an update rule depending on the firing time of only two neurons; they result in interesting patterns of global behaviour which includes the competition between synapses (Van Rossum et al. 2000). Despite many important features of the Hebb's rule, there are also some drawbacks associated with it. The first drawback is causality such as in Hebb's rule the firing of neuron A must precede the firing of neuron B. So Hebb's rule is not valid for the simple correlation in which the order of spike times is unimportant. Hebb's discussion only considers neural interaction mediated by spikes, so the critical role is explicit for the postsynaptic cell and implicit for the presynaptic cell. Finally Hebb's rule only considers the conditions under which the synaptic efficacy increases; it doesn't describe the condition under which the synaptic efficacy decreases. Since the information in the nervous system is coded by spike rate instead of timing of individual spikes, the central roles of causality and spike-timing have been under scrutiny (Roberts and Bell 2002). Our system includes aspects of these ideas by allowing the variation of three different types of neural weights in response to the firing of neurons.

5.3 Short Term Memory

Eric Kandel (2000 Nobel Prize winner), in 1965 discovered that the mechanism for short term learning is situated at the synapse. His discovery was made possible through his studies on the marine snail *aplysia* in habituation, sensitization and classical conditioning. Kandel traced the electrophysiological changes which are caused by these combined stimuli to specific synapses (Kandel et al. 2000).

Kleim showed that the rats could develop enhanced synaptogenesis when exposed to motor skill learning tasks (Kleim et al. 1998). It has also been shown that animals which live in complex environments can develop greater dendritic branching. These variations can be developed very quickly and their duration can range from minutes to hours (Greenough et al. 1985).

5.4 Long Term Potentiation (LTP)

The increase in synaptic strength between two neurons which is caused by their simultaneous stimulation is called LTP. LTP is considered to be one of the integral mechanisms which are responsible for the memory in the brain, since the learning experiences are all stored in the form of synaptic strength between the neurons (Cajal

1894). Memories are not dependent upon neural growth; they might be formed by improving the effectiveness of the communication between the neurons through strengthening the connections between the existing neurons.

Terje Lomo discovered the LTP for the first time in 1966 while working in the laboratory of Per Andersen (Terje 2003). He conducted several neurophysiological experiments on the anesthetized rabbits for exploring the role of the hippocampus and the process of LTP.

LTP increases the sensitivity of post synaptic neurons to the signals received from presynaptic neurons; hence it enhances the communication between neurons as well (Malenka and Bear 2004). Increasing the activity of existing receptors or increasing the number of receptor on the postsynaptic neurons can improve the sensitivity of the post synaptic neurons. This results in a permanent variation of synaptic signalling which results in long lasting effects and is considered to be responsible for production of long term memory of the event.

5.5 Developmental Plasticity: Synaptic Pruning

The process in which the weaker synaptic contacts are eliminated, while the stronger connections are preserved and strengthened is known as synaptic pruning. At times, during the developmental phase; a greater number of structural elements such as neurons and neurites are produced than the actual requirement. This is termed as an overshoot phenomenon (Ooyen and van Pelt 1994). Due to the overshoot, the system can then tune itself soundly on the basis of environmental conditions (GoodMan and Shatz 1993). The death of inappropriately connected neurons or the reduction in the number of synapses maintained by the individual neurons or both can be used for refinement.

The strengthening and pruning of connections can be determined through experience. The connections which are activated more frequently are preserved. There must be a purpose for a neuron to live; else it will die through a process called "apoptosis". In this process, neurons which neither receive nor transmit information are damaged and then die (Gopnic et al. 1999). The ineffective or weak connections are pruned. The plasticity is responsible for the process of developing and pruning connections; thus allowing the brain to adapt to the environment.

Many aspects of neuroscience which are relevant to the model were discussed in this chapter. The building block of brain are neurons. They can arrange themselves in variety of structures. Based on these structures they attain their functionalities. The model discussed in this book includes the neural morphology as it allows the neurons to have random number of dendrites, dendrite branches, and axon branches. It also allows the branches to grow and shrink along with the variation in architecture in response to the environmental signals. There is an artificial space given to the neurons in this model which assists them in interacting with their neighbours through making synapses. The neuron in the model discussed in this book has three main parts which are the soma (cell body), dendrites with its branches, and an axon with its branches. The potentials of all the branches and the soma after every cycle are reduced, in an attempt to replicate the biological neuron. A mechanism in which the signals from all the neurons in the environment are received through dendrites is also implemented, and then the decision about firing or not firing the action potential is made. This is also inspired from the process of action potential described in Sect. 6 (Electrical Signalling). Synaptic Plasticity lays the foundation for learning and memory development in the brain as mentioned in Sect. 2.7. Using it as the guideline, we have also implemented a mechanism synaptic plasticity in the model of this book's network as well through the addition of weight to each branch and soma updated at runtime during the development.

The next chapter provides detailed discussion on evolutionary computation. It explains all the methods relevant to the work of this book, in detail.

Chapter 3
Evolutionary Computation

This chapter is divided into four main sections.

- Evolutionary Computation (EC)
- Cartesian Genetic Programming (CGP)
- Co-Evolutionary Computation (CC)
- Developmental Systems (DS)

1 Evolutionary Computation

The field of Evolutionary Computation has existed for more than 50 years (Fogel 1998). The basic idea behind the evolutionary computation is that the candidate solutions of problems can progressively become better through some combinations of mutation (varying a solution randomly) and cross over (mixing two solution to form a new one) to vary the solution as evolution continues. A fitness function is used to measure the candidate solutions that can give an approximation of the closeness of the proposed solution to the desired solution.

Evolutionary Computation can be divided into four main categories:

- Evolutionary Strategies (ES)
- Evolutionary Programming (EP)
- Genetic Algorithms (GA)
- Genetic Programming (GP)

A brief discussion of these types are presented below.

G.M. Khan, *Evolution of Artificial Neural Development*, Studies in Computational Intelligence 725, https://doi.org/10.1007/978-3-319-67466-7_3

1.1 Evolutionary Strategies

They were first introduced in Germany by Rechenberg (Rechenberg 1971) and later developed by Schwefel (Back et al. 1991). The purpose behind their design was to find the function minima for large multi-variant functions and discover a real valued vector that can minimize a function for the problem. The earliest algorithm relied on a population consisted of a parent and a child. The child arose from the parent through the addition of normally distributed random numbers to the elements of the vector. The child would then be evaluated for its performance to be better than the parent and allowed to replace the parent at the next iteration. This was called the two membered ES.

Later on Rechenberg proposed a multi-membered ES. The multi-membered ES consisted of more than one parent (μ) and child (λ) (Rechenberg 1994). This type of ES was represented by μ+1 ES. It was further improved by increasing the number of off-springs. Then the new ES was called μ+λ ES. This selection operates on both parents and off-springs. The parents survived until better off-springs were produced. This technique when applied to problems involving dynamic fitness functions would often get stuck at a non-optimal position. To solve this problem, Schwefel introduced a generational ES known as (μ, λ) ES. In this case every parent was replaced by the offspring in the upcoming generation. The system presented in this book is based on 1+λ ES which indicated the presence of one parent and more offspring. Miller has shown that this algorithm works well for many problems in CGP (Miller 1999; Miller et al. 2000).

1.2 Evolutionary Programming

Evolutionary Programming was first introduced in 1966, when presented as candidate solutions for finite state machines to evolve them (Fogel et al. 1966). Earlier work was based on population of three and five solutions, and the genetic alteration was carried out by one of the five types of mutations as mentioned below.

(1) Adding a state and randomly assigning all transitions for that state.
(2) Deleting a state and randomly assigning any transitions which were previously feeding into that state.
(3) Change an output symbol on a transition.
(4) Alter a transition associated with an input symbol.
(5) Change the starting state.

Mutation is employed as the primary variation operator. Rather than being considered as members of the same specie, each constituent of the population is reckoned as a part of specific specie. Every member of population can produce its own off-spring. Just like ES and GA's, EP is a useful method of optimization when techniques like gradient descent or analytical discovery are not possible.

1.3 Genetic Algorithms (GAs)

The GAs were first invented at University of Michigan by John Holland. Further development took place as a group of Holland, his students and colleagues at the University of Michigan in 1960s and 1970s (Holland 1975a). Holland's method involved the study of the natural phenomenon of adaptation, and then using it as a guideline; this mechanism could have been imported into the computer systems. They are often used for optimization issues, where the form of solution is already known. A fixed length binary string can be used to find the variable involved in finding the solution. The algorithm starts with a "population" of randomly generated possible solutions of the problem which are then allowed to evolve over generations for better solutions. The population is actually a collection of the candidate solutions which are considered during the course of the algorithm. Each solution in the population is considered to be an individual. Generation after generation, new members are added and the older members are removed from the entire population. A number termed as fitness is found out for every individual which indicates how good the solution is. The selection of better individual is done probabilistically. The individuals of better caliber are then used to produce new population, either through recombination or through mutation. The mixing of the two solutions result in recombination or 'cross over' for the production of new individuals. By slight variation to each individual, we can achieve mutation as well. The size of population can highly vary. The number of possible solutions is dependent upon the population size, and if the number of possible solutions is more; then there is a higher variation in the population. Variations are indication of better solutions being created. That is why there is a need for increasing the population as much as possible. However, the increase in population cause an increase in the computational burden. Thus an optimal size of population based on the type of problems is always challenging.

1.4 Genetic Programming

It is a technique to achieve "computer programs" that are able to solve user defined tasks by using evolutionary strategies. Nils Aall Barricelli in 1954 laid the foundations for GP through evolutionary algorithms (Barricelli 1954). However the work of Ingo Rechenberg in the 1960s and early 1970s made evolutionary algorithms as a popular optimization method (Rechenberg 1971). The work of John Holland was also very influential in the 1970s (Holland 1975b). Stephen F. Smith and Nichael L. Cramer reported the first result on the GP methodology (Smith 1980; Cramer 1985). Forsyth put forward evolution of small programs in forensic science for the UK police in the year 1981 (Forsyth 1981). Schmidhuber proposed Meta-GP which is a form of GP (Schmidhuber 1987). It is a recursive algorithm which can be stopped, resulting in avoiding an infinite recursion. The major aim of his work was to produce a GP system which could work on the principle of finding a better program modifying program which means a GP for improvement of GP or learning for learning. John R. Koza is the main proponent of GP and has founded the application of genetic programming in various complex optimizations and search problems (Koza 1990, 1992, 1994; Koza et al. 1999, 2003). Following the AI computer language, LISP, Koza used trees to implement programs. The terminals functioned as the inputs of the program along with some constants. If the desired output is either a number or a list of numbers, the common choice for the function would be the standard arithmetic and transcendental operators like +, −, exp, and log. If the output is in form of Boolean or logical value, then the choice for function would be logical operators like AND, OR, and NOT. One can also use some other appropriate functions for other problems.

Usually the generation of random individuals, takes place through a recursive algorithm which arbitrarily selects a non-terminal or end stage at each node. The selection of terminal concludes in termination of the recursion. However, the selection of a non-terminal results in the application of algorithm recursively for generating children nodes. The maximum size of the tree is defined, and those trees which have greater size then the maximum defined size; they are not generated. The maximum size might be maintained throughout the entire process. Mutation and cross over will result in the production of new individuals. A parental individual is selected through the probability based on fitness and then mutation operates on this parental individual. Any node of the tree can be chosen randomly and then mutated for the production of new individuals. Mutation is usually applied at lower rates. For the sake of crossover, two parental individuals are selected through the probability based on fitness. One node is chosen arbitrarily from the tree of the individual parent. Then the sub-tree based at the selected node in one parent is swapped with the sub-tree based at the selected node of the other parent. Both mutations and crossovers are applied separately which means that either crossover, mutation or none is applied to each individual. Crossover and mutation cannot be applied to the same individual. For obtaining the fitness of the individual, the program is tested on different test cases. On the basis of the selection criteria, the best individuals are promoted to the next generation.

In this section, we discussed the evolutionary computation along with its relevant sub fields. We utilized the evolutionary strategies in our computational network and have also used a special form of GP known as the CGP (explained in the next section) for implementing a genotype of the model discussed at a later stage.

2 Cartesian Genetic Programming (CGP)

Miller and Thomson's work laid the foundations for the Cartesian Genetic Programming while evolving feed forward digital circuits (Miller et al. 1997; Miller 1999; Miller and Thomson 2000). Directed acyclic graphs are used to represent the programs in CGP. The advantages of these graphs are that they can allow the re-use of sub graphs. The original form of CGP used a rectangular grid of computational nodes where the nodes were not allowed to take their inputs from a node of the same column. However, with the passage of time; this restriction vanished by letting the number of rows always selected to be one (as used in the model of this book). The genotype in CGP has a fixed length. The genes are integers which can encode the function and connections of each node in the directed graph. The phenotype can be obtained through following referenced links in the graph which might indicate that some of the genes are not referenced in the path from the inputs of the program to its outputs. The result is a variable length bounded phenotype. This might cause some non-coding genes which do not influence the phenotype at all; thus there is a neutral effect on genotype fitness. Characteristics of such genotypic redundancy have been investigated thoroughly and it has been extremely beneficial to the evolutionary process on the problems studied (Miller et al. 2000; Vassilev and Miller 2000; Yu and Miller 2001, 2002).

Every node in the graph is the representation of a particular function and its connections. One gene is used for encoding the function of the node, while the remaining genes used for encoding the input connections. The inputs of the nodes are either the output of the previous node or a program input (terminal). The number of inputs to the node is dependent on the number of inputs required by the function it is representing.

Recent works have brought up the module acquisition and evolution into CGP (Walker and Miller 2004). This also showed that these techniques are even more scalable on harder problems. Our work in this book hasn't taken these methods into account.

CGP is used to find the unknown neural functions and neural developmental programs inside the neurons of the model presented in this book.

3 Co-Evolution

The result of two species affecting each other's evolution is termed as co-evolution. It might be either cooperative or competitive. This is a not a phenomenon which occurs through the environment, indeed it occurs through one-on-one interaction; such as predator and prey, host-symbiont or host-parasite pair. Competitive environment is the platform where the maximum use of Co-evolutionary computation takes place (Pollack et al. 1996; Rosin 1997). The interactions are either between individuals who are competing in a game context or between the different populations which are competing in predator-prey type relationships (Hillis 1991; Paredis 1994a, b; Cliff and Miller 1996; Juille and Pollack 1998).

An individual's fitness in a competitive coevolution is evaluated on the basis of how the individual performs against the opponent in the population. However, fitness is a mere indication of the comparative robustness of the solution; not the absolute strength of the solutions. This results in a relative decrease in the fitness of the opponent. These competing solutions lead to an 'Arms Race' for better solutions (Dawkins and Krebs 1979; Van Valin 1973). The feedback mechanism among each individual on the grounds of their selection push them towards increased complexity.

Cooperative coevolution has been evaluated to solve different problems such as the difficulty in choosing a proper encoding mechanism for the individuals and the difficulty in decomposing the composite problems (Paredis 1995; Jong and Potter 1995). Studies have shown that there is a need for balanced cooperation and competition for preventing evolutionary algorithms from being stuck in local optimum, or mediocre stable states (Ficici and Pollack 1998).

Predator-Prey relation is one of the best examples for co-evolution. The task of the prey is to defend itself against the predator by developing new strategies such as running quicker, growing bigger shields, and better camouflage in response to the predator's action. The predators also respond to the improvement in their prey by improving their attacking strategies such as stronger claws, and better eyesight. This battle of improvement goes on between the prey and predator due to which they adapt to complex strategies. Hillis was the first one to come up with the idea of a co-evolutionary competitive learning environment of predator and prey where both of them strive to accomplish their assignments and persist (Hillis 1990). Traditionally, Competitive evolution is utilized for evolving interactive behaviours and are intricate to develop with absolute fitness function. Karl Sims evolved 3D creatures in a simulated environment, where the goal was to capture the ball before the opponent captures it (Sims 1994). This resulted in various effective interactive strategies.

Co-evolutionary algorithms are of great importance in artificial life, optimization, game learning and machine learning problems. The past several years have witnessed the use of evolutionary techniques in various games such as Othello (Moriarty and Miikulainen 1995), Go (Lubberts and Miikkulainen 2001), Chess (Kendall and Whitwell 2001), Kala (Irving and Uiterwijk 2000), (Wee-Chong and Yew-Jin 2003) and Checkers (Schaeffer 1996), (Fogel 2002). Chellapilla and Fogel utilized co-evolution of Artificial Neural Networks (ANNs) and were able to evolve an ANN which could play at a master's level. Lindgren co-evolved iterated prisoners Dilemma strategies in order to exhibit their correspondence to phases in natural evolution (Lindgren and Johansson 2001).

A new method called Pareto Co-evolution has already been introduced by Ficici and Noble which selects the unrivaled apprentice and mentors of the two populations by using co-evolution as multi-objective optimization problem (Ficici and Pollack 2001; Noble and Watson 2001). This allows the best individuals to reproduce, while maintaining an informative and diverse set of opponents. If the search space is fixed then it will not be open ended and will stop at local optimum. It is not dependent upon how well the selection is performed or how well the competitors are chosen. At times, the addition of a new search space can assist in escaping from a local optimum which is relatively easier compared to searching for a new path through the original space. Such techniques which increase the complexity can be used to add new dimensions to the search space. The process continue indefinitely, even if the global optimum is attained while searching space of solutions. The addition of new dimensions can open up higher dimensional space where even better optima might exist.

The ability of a network to play against a high quality player of possible opponents encourages interesting and sophisticated strategies in competitive co-evolution (Stanley and Miikkulainen 2004). This can be achieved through the evolution of two separate populations, where each population is evaluated against an intelligently chosen sample of networks from the opposing population. The population which is evaluated is called the host population while the opponent's population is called the parasite population (Rosin and Belew 1997). This strategy makes evolution more efficient and more reliable compared to the one based on random or round robin tournament.

Floreano and Nolfi used a master game approach where the champion of each generation was compared with the champions of all the other generations (Floreano and Nolfi 1997). However, this technique is expensive, time consuming and a huge number of resources are required to undergo all these competitions. This section presented a detailed explanation of co-evolution through different applications and methods tried. The co-evolutionary strategies have also been used in the Wumpus world and checkers problem (later in the book). Co-evolution was found helpful in generating learning capabilities in the networks.

Next section presents a detailed explanation of development and developmental systems.

4 Developmental Systems

In the biological world, the complex systems are formed from simple gene structures through developmental processes. The same process can be used in computation development for the production of complex systems from simple systems capable of learning and adaptation.

There are some critics of development that believe that as long as evolutionary algorithms can solve the problems, there is no need for development (Kumar 2003). But there is a serious problem with the use of traditional evolutionary algorithm, which is that the genotype usually has a one to one relationship with the solution description; and with an increase in complexity of the problem; the size of genome which encodes the solution increases with the same ratio. If the solution to a problem has n constituents, then there is a dire need for at least n genes. But the value of n can go as high as thousands or millions. To solve this problem, we can add constraints as rules and or knowledge in the genes and allow the system to develop on the basis of these rules. This will result in the development of a complex network from a simple genotype, which can solve different problems. The biological kingdom ranges from microscopic to gigantic organisms; however they all use similar fundamental process of development. Due to this development, evolution has been able to give rise to greater complexity.

Evolutionary algorithms rely on the use of a direct and linear calibration between the genotype and phenotype for evolving outcomes in computers, while development relies on highly indirect and non-linear mappings. According to Kumar and Bent (Kumar 2003) (pp. 13–14), "for development to occur in computer models, we require a developmental encoding, i.e., our genes must act like instructions. Development then becomes the process (es) of executing those instructions and dealing with the highly parallel interactions between them and the structure they create."

The above definition clearly identifies the method of computation development capable of solving problems. The genes must have the information of developing a system. In the biological system the genome has a complete description of the phenotype and does not have any direct mapping with the developed phenotype and its functionality. The genome has no interaction with the environment, so it develops a phenotype which can interact with the environment. During the development, the Genes interact with each other at different hierarchical levels in the phenotype. This results in the ability of self-organization, which is similar to complexity developed in an ant colony (Holland 1998; Bentley 2002). Evolutionary algorithms can get stuck at local optima. There are measures such as increasing mutation rates which can counter it, but they are useful only at early stages of evolution when solutions are not very fit. The small mutational variations of developmental systems can sometimes completely change the kind of phenotype developed, which might result in the added problem of being trapped (Kumar 2003).

The above section explained the development in general. The next chapter will present detailed developmental models of neural networks. The introduction of developmental genetic programming in the model discussed in this book is for controlling the developmental process or neural structures on the task environment (Chap. 5 contains the details).

This chapter presented a review of Evolutionary Computation, Cartesian Genetic Programming, Coevolution and Developmental Systems. The chapter presented an overview of Evolutionary Computation along with its subfields consisting of Evolutionary strategies, Evolutionary Programming, Genetic Algorithms and Genetic Programming. The reason behind the review of the Evolutionary Strategies was that they are used as the selection process in the evolution of the system (Cartesian Genetic Programming Developmental Network (CGPDN)) as mentioned in Sect. 2.2 in Chap. 5. The unknown functions of CGP neurons which help in their development and electrical processing were found out through the Cartesian Genetic Programming (Chap. 5 Sect. 5.3 contains the details). The CGPDN system has been tested in a co-evolutionary environment in Wumpus world (Explained in Chap. 6) and checkers (Explained in Chap. 7) environments. Co-evolutionary behaviour has been described in Sect. 2, Chap. 6. The CGPDN develops a complex neural structure (see Chap. 5) while solving a particular task.

The purpose behind the use of CGP is that it is a highly effective form of GP and its hardware implementation is simple and convenient (Walker and Miller 2008). The conventional ANNs could have also been used in place of CGP for finding out the unknown functions inside the neuron, but their hardware implementation is much more complex; since ANNs use floating point numbers and non-linear mathematical functions such as sigmoid or hyperbolic tangent. The next chapter presents a detailed description of the Artificial Neural Network, with description of the background and related work in the field, neuro evolution, neural development and the major problem with ANNs, 'The Catastrophic Forgetting'.

Chapter 4
Artificial Neural Network (ANNs)

This chapter presents a review of the major forms of the Artificial Neural Networks (ANNs) (Sordo 2002). The particular topic of discussion of this chapter is how learning takes place in these models. Different ways of training the networks are examined.

- Background of the ANNs, its structure and applications
- Kohonen Self organizing maps and Hop-field networks
- Historical perspective of ANNs and its Evolution
- Applications and importance of Computational Development in the field of ANNs
- Catastrophic Forgetting
- Conclusion and summary of the relevance to the CGPDN.

1 Artificial Neural Network

The computational systems made up of interconnected neurons are termed as Artificial Neural Networks (ANNs). The properties of these neurons resemble those of the biological neurons. They can exhibit complex global behaviour which is dependent upon the interconnection of neurons, their internal parameters and their functions. These artificial neurons are bound together through different connections. The seamless transmission of signals from one neuron to another neuron takes place through these connections.

G.M. Khan, *Evolution of Artificial Neural Development*, Studies in Computational Intelligence 725, https://doi.org/10.1007/978-3-319-67466-7_4

1.1 Applications of ANN

ANNs are used in different real life applications such as function approximation, time series prediction, classification, sequence recognition, data processing, filtering, clustering, blind signal separation, compression, system identification and control, pattern recognition, medical diagnosis, financial applications, data mining, visualization and email spam filtering (Dorffner and Porenta 1994; Dorffner 1996; Sjoberg et al. 1994; Timothy 1994; Murray 1993; Ripley 1996).

1.2 History of ANN

The first generation of Artificial Neural Networks were based on the McCulloch-Pitts threshold neurons, which generated binary outputs (McCulloch and Pitts 1943). If the weighted sum of the inputs is above the threshold value, the unit was taken as 'on'; else the unit was taken as 'off'. The nature of inputs is either decimal or floating point numbers. The output of these neurons is only digital, but they have been successfully applied in artificial neural networks. The second generation Neurons utilize a continuous activation function for calculating their output. It makes them suitable for analogue input and output. Some of the frequently employed activation functions are sigmoid, and hyperbolic tangent. The second generation neurons are regarded as stronger than the first generation neurons. If the output layer of the second generation uses first generation binary units, they can be used for digital computations with few neurons in comparison to a network consisting of only the first generation units. They can also be used to approximate any analogue function which makes these networks universal for analogue computation (Maass et al. 1991). The continuous output of second generation ANN units can be interpreted in terms of a firing rate model. This value indicates the normalized firing rate of the unit in response to a particular input pattern. That is why second generation neuron models are considered a close approximation to the biological neurons and they are also more powerful than the neuron models of the first generation (DasGupta and Schnitger 1992).

The third generation of neural networks generate individual output spikes; hence they are closer to biological neurons (Ferster and Spruston 1995). The outputs can be interpreted using pulse coding mechanisms. The neurons send and receive individual pulses. The third generation networks are sometimes termed as Spiking Neural Network (SNN) (Gerstner and Kistler 2002) as explained in the next subsection. A wider range of neural coding mechanisms are entertained such as pulse coding, rate coding and mixtures of the two (Gerstner et al. 1999).

The recent experimental results have shown that the neurons of the cortex can carry out analogue computations at a very high speed. It has also been shown that the human's analysis and classification of visual inputs take place under 100 ms (Thorpe et al. 2001). At least 10 synaptic steps are required from the retina to temporal lobe, thus leaving 10 ms of processing time per neuron. This time is considered to be too short for averaging mechanisms like rate coding for processing the information. So whenever speed is an issue, pulse coding schemes are thought to be the best (Gerstner et al. 1999; Thorpe et al. 2001).

1.3 Spiking Neural Networks (SNN)

The interaction between the biological neurons take place through a short pulse called action potential or spikes (Gerstner and Kistler 2002). Recently, researchers have shown that neuron can encode information in timing of these spikes instead of average firing frequency. The implementation of these SNN models takes place on this principal. Both in conventional ANNs and Spiking Neural Networks (SNNs), the information is usually distributed in the weight matrix. The interval between the time of spike of post synaptic neuron and pre synaptic neuron is used to adjust the weights in the SNN. The processing of a rapid temporally changing stimulus, which cannot be reproduced by having more neurons or connections, is only possible through the synaptic plasticity (Mehrtash et al. 2003).

1.4 Mode of Operation

The artificial networks can operate either in a learning (training) or testing mode. Once the learning starts, from a random set of parameters, the weights and thresholds are continuously updated until the desired solution is obtained; the parameters are then frozen and remain fixed during the testing process. During the adaptive process of learning, the weights between all the interconnected neurons are updated until an optimum point is attained. The weights of the network can be either floating point numbers or parameter dependent functions.

1.5 Learning Rules

The methods used for adjusting certain quantities responsible for the learnt information, typically weights are termed as learning rules. Supervised and unsupervised learning are the two main mechanisms of learning. When a desired output result is used to guide the update in the neural parameters, it is termed as supervised learning. While in the later mechanism, the training of the network is entirely dependent upon the input data and there is no provision of the target results for updating the network parameters which can be used to extract features from the input data (Hinton et al. 2006).

Back-propagation and evolutionary methods are the two conventional learning methods. In the back-propagation, the output and the desired results are compared with each other, and the error is reflected backward to update the weights of ANNs accordingly. In evolutionary methods, the weights of the best performing ANN are slightly changed (either through mutation or cross over) for the production of next set of weights. In this manner, the optimum performance weights are obtained. Back propagation is also used for multilayer perceptron having input layer, hidden layers and output layer. Cost function is the predefined error function which can be calculated by comparing the output with target in back propagation. The cost function is given by

$e = f(d_i - y_i)$

Where;

d_i = The desired value

y_i = The system output

e = error

In order to reduce the error, it is fed back such that the weight of each connection is adjusted in a direction that minimizes the overall error. The process is repeated for converging the network to a state of minimum possible error.

Gradient Descent is an optimization method used for adjusting the weights in a manner which reduces the net error. The error function is differentiated with respect to the network weights. On the basis of the results of differentiation, the weights are adjusted for reducing the error. Because of this reason, back propagation is applied to networks which have differentiable active function.

The units of the intermediate layer of the feed forward neural network can be instructed through back propagation algorithm. The features of the input vector for predicting the desired output are represented by these units (Rumelhart et al. 1986). This training can be performed through the provision of information regarding the discrepancy between the actual output and the desired output of the network in order to customize the connection weights to reduce the discrepancy.

2 Types of Neural Networks

Different types of neural networks have been introduced over the years, but the most common one are feed-forward, Kohonen Self organizing maps, and Hopfield neural networks.

2.1 Feed Forward Neural Network

These networks are usually arranged in the form of layers where each layer has a number of neurons as the processing units (Sordo 2002). Signals are transferred from layer to layer through the input-output manner, where signals are processed at each layer and transferred in the forward direction. This basic architecture of a traditional ANN is called Multilayer Perceptron (MLP). Figure 1 presents a sketch of a general

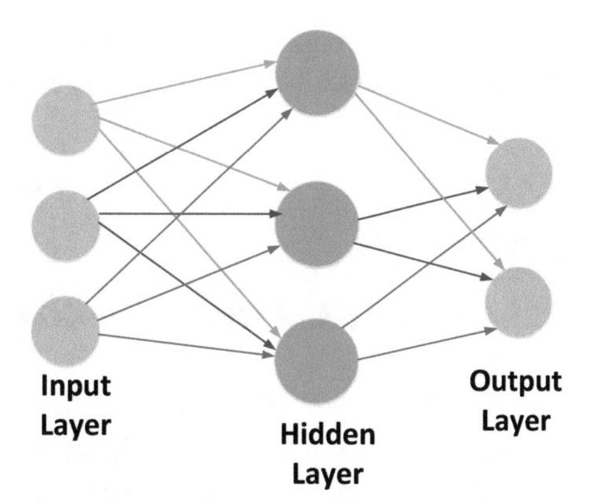

Fig. 1 Multilayer perceptron

model of Multi-layer perceptron consisting of an input layer, hidden layer and an output layer.

There are usually two layers of processing elements and a hidden layer in the MLP networks (as shown in Fig. 1), however the number of hidden layers can vary. The external signal arrives at the input layer which is then propagated by the input layer to the next hidden layer as a weighted sum. The hidden layer processes it through the activation function. The commonly used activation functions are hyperbolic tangent, the value of which ranges from −1 to 1;

$$\phi(x_i) = \tanh(x_i)$$

and sigmoid function with values range from 0 to 1;

$$\phi(x_i) = (1 + e^{-x_i})^{-1}$$

x_i is the received weighted and summed up signal from the input layer.

The job of the hidden layer is to transfer the processed signal to the neurons of another hidden layer, and if it is the last hidden layer; then it transfers the processed signal to the output. The signals reach output as the weighted sum, processed through the activation function. The output of the network is taken from the last layer.

The training of MLP networks is carried out by altering their connection weights after every processing interval. The variation in the weights is dependent upon the error between the output and the desired value. Usually this is done through back propagation. The error *(e)* in output node *j* in the *nth* data point is given by;

$$e_j(n) = d_j(n) - y_j(n)$$

where;

d = target value

y = value produced by the perceptron (Haykin 1998).

The error can be used to adjust the weights of the nodes in a manner that the energy $\mathcal{E}$ of error in the entire output is minimized as given by:

$$\mathcal{E}(n) = \frac{1}{2}\sum_{j} e_j^2(n).$$

2.2 *Kohonen Self Organizing Neural Networks*

The Self Organizing Maps (SOMs) which are used as computational methods for the visualization and analysis of high dimensional data were introduced by Teuvo Kohonen (Kohonen 1982, 2001). The maps are based on unsupervised competitive learning whose source of inspiration is the biological structure of the cortex. Cortex has different areas which are responsible for different human activities (motor, sensory, visual and somatosensory). Every sensory area is mapped to the corresponding area in the cerebral cortex. It is thought that the cortex contains the self-organizing computational map of the body. The sensory cortex also preserves the spatial relations between the body parts as much as possible. The same phenomenon also occurs in the motor cortex.

The self-organizing networks have a two layer topology (as shown in Fig. 2). The first layer is the Input layer while the second one is the Kohonen Layer. There is a node for each dimension of the input in the input layer where every input is connected to all the nodes in the Kohonen layer hence the two layers are fully connected. The node value in the Kohonen layer represents the output. The number of nodes in the Kohonen or output layer must be at least equal to the number of categories to be recognized. One neuron in the output layer has to be activated for every dimension of the input. The Kohonen layer neurons are neighboured by the grid (Kohonen and Somervuo 2002; Kaski et al. 1998; Martinetz et al. 1993). These networks are of great importance in applications, such as data clustering which occurs in speech recognition and handwriting recognition for sparsely distributed data. Lateral inhibitions are used by them which are inspired by the vision system working in biological neural systems.

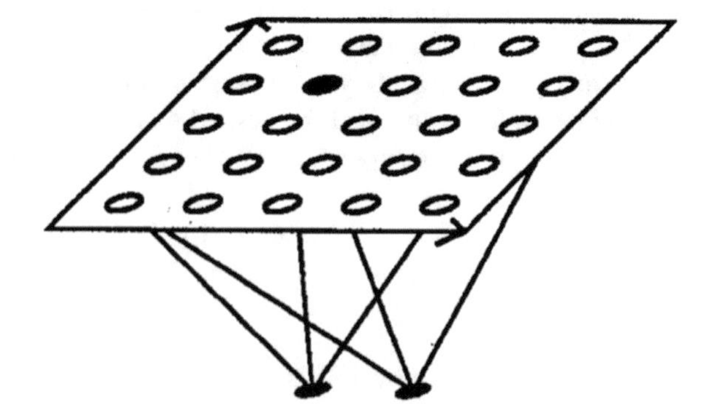

Fig. 2 Structure of Kohonen Self Organizing map, showing input neurons and the kohonen layer neurons. Input neurons are fully connected with the kohonen layer neurons, A winning neuron represented by a black dot. Taken from (Hertz et al. 1991)

Kohonen networks rely on the principal of mapping input vectors (pattern) of arbitrary dimension onto the Kohonen network in a way where the sequences closer to each other in the input space should be within close range. The training in Kohonen network begins with the fairly large sized neighbourhood of the winner. As the training proceeds, the distance reduces. The unit whose weight vector has the shortest Euclidean distance from the input sequence is the winning output unit. The neighbourhood of a unit consists of all the units which lie in its proximity on the map (not in the weight space). In the process of training, the closely distant node is selected along with its neighbour's weight; the modification for increasing the similarity with input takes place. The radius of neighbourhood decreases with the passage of time, and finally only a specific area in the network is identified for an input pattern is left. The following equation is used for updating the weights of the winning unit along with its neighbourhood.

$$w_i = w_i + \alpha(x_i - w_i)$$

where;

w_i = The weight of ith unit

x_i = The input

α = The Kohonen's rule for adjustment of weights.

In order to model the directional motion in the visual cortex, Farkas and Miikkulainen used SOMs (Farkas 1999). Their neuron model has 'leaky integrators' at synapses. It performs time-dependent summation with decay of incoming spikes. Once the dynamic threshold is exceeded, then a spike is fired. The spikes decay exponentially with time and are accumulated over a set of afferent and lateral inputs. The weighted output from leaky integrator is then applied to the spike generator. The spike generator will generate a spike only if the input threshold is exceeded. The output spike is then applied again for increasing the threshold, which makes it less likely to produce the second spike. There is an exponential decrease in the threshold with time. Every node has the receptive field of the receptors in the retina. They are weighted and integrated over time for creating a Hebbian type weight adjustment.

2.3 *Hopfield Networks*

A recurrent neural network is known as Hopfield Network (Hopfield 1982). Recurrent networks possess the property of bi-directional flow of information i.e. forward and backward direction. The nodes in such networks are fully connected to each other and they can function as both the input and output. The idea behind it is that the instability of states is iterated until a stable state is attained. This guarantees the convergence of the dynamics (Fig. 3).

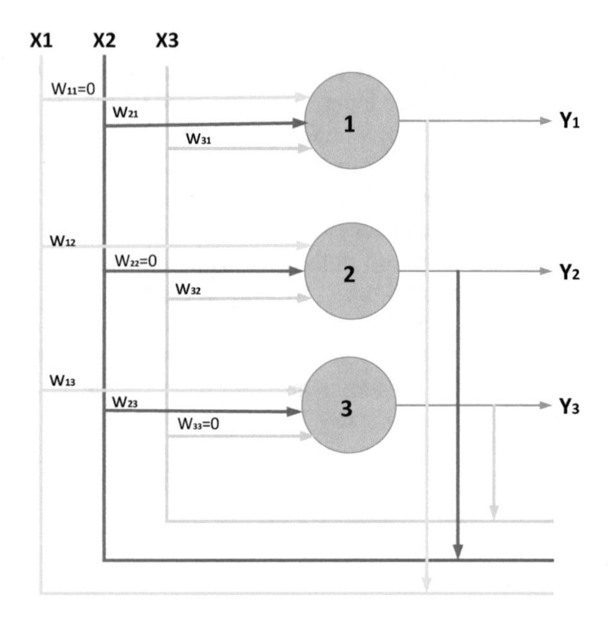

Fig. 3 Hopfield Network: Three node Hopfield network, with x_i = Input, y_i = Output, and w_{ij} = Weights attached to connections

The processing units which are used are binary threshold units. The binary threshold units only take two different values for their state. The values can be either -1 and 1, or 1 and 0. The two possible definitions for the activation y_i of unit i's are

$$y_i \leftarrow \begin{cases} 1 & \text{if } \sum_j w_{ij} x_j > \theta_i, \\ -1 & \text{otherwise.} \end{cases} \tag{1}$$

$$y_i \leftarrow \begin{cases} 1 & \text{if } \sum_j w_{ij} x_j > \theta_i, \\ 0 & \text{otherwise.} \end{cases} \tag{2}$$

where;

w_{ij} = weight of the connection

x_j = The state of unit j

θ_i = The threshold of unit i

There are two main restrictions in the connections of Hopfield net.

- No unit must be connected with itself
- Connections are symmetric

There is an energy function associated with every state of the network in Hopfield net given by the following equation:

$$E = -\frac{1}{2} \sum_{i \neq j} w_{ij} x_i x_j + \sum_i \theta_i \, x_i$$

where;
E = Energy of the network state.
w_{ij} = The weight of connection.
x_i = The state of unit i
x_j = The state of unit j
θ_i = The threshold of unit i

With update in the network, there is decrease in the energy till a minimum is reached. In Hopfield networks, the energy of the states is lowered in the training phase. However, the network keeps track of its energy state (Hopfield and Tank 1985). The network can approach a previous state if it is granted only a portion of that state, hence working as a content addressable memory system. We can also recover a distorted input from the trained state of the network. The input most similar to the distorted form is used as the recovery. As the memory recovery is based on the basis of similarity, therefore it is termed as the associative memory. As a result, Hopfield Networks are sometimes called associative networks as well.

Hopfield networks can be used in many optimization problems. The problem first has to be transformed into variables in a way that the desired optimization corresponds to the minimization of the respective energy function (Hopfield and Tank 1985). Hopfield networks can also be applied to the non-linear factorization problems (Husek et al. 2002).

In the next section, inclusion of artificial evolution into ANNs is described.

3 Neuro-Evolution

This section describes the Neuro-evolution (NE), which involves the use of artificial evolution with ANNs. Neuro-evolution refers to the evolution of various aspects of neural network. It is a combination of ANN and genetic algorithm, with ANN being the phenotype and genetic algorithm being the corresponding genotype. The genotype can represents the connection weights, connection type, node function, topology of ANNs or combination of any two, three or all the parameters. The genotype is evolved until desired phenotypic behaviour is achieved. Encoding is the important aspect of the NE system, since it affects the search space of the solutions (Yao 1999). Depending upon the methods, either the weights of the network or the topology or both are evolved. When a fixed topology network is used and only weights are evolved, the network solution space is constrained; it has to work in a restrictive environment not attaining any novel solution to the problem. It is not an easy job to select the proper topology of ANN for a specific problem. The Topology and Weight Evolving Neural Networks (TWEANNs) evolve both weights and network topologies. In this method, evolution is provided with flexibility for selecting the desired topology and weights for its network. So, TWEANNs genotype can encode both the topology and weight of the network. This increases the efficiency of the network, but it comes at the cost of increase in computational cost.

TWEANNs can also use both direct and indirect encoding methods of genotype. In direct encoding of genotype, every connection and node in the phenotype has to be specified in the genotype (Zhang and Muhlenbein 1993; Lee and Kim 1996; Dasgupta and McGregor 1992; Opitz and Shavlik 1997; Yao and Liu 1996; Angeline et al. 1993; Maniezzo 1994). In indirect encoding, only the rules for constructing the phenotype are specified in the genotype (Bongard and Pfeifer 2001; Gruau et al. 1996; Hornby and Pollack 2002; Mandischer 1993). The genotype doesn't specify every node and connection in the phenotype in the indirect encoding. TWEANNs which utilize the indirect encoding use a developmental approach that is akin to an artificial embrogeny (AE) (Stanley and Miikkulainen 2003) in which the small phenotypical structures act as the starting point which are developed to produce the final phenotype.

ENZO (Evolver and Network optimizer) is a system which can optimize both topology and connections' weights at the same time. ENZO uses direct encoding scheme (Braun and Weisbrod 1993). The set of the possible connections is fixed as the gene corresponds to a connection in the network. ENZO scheme provide introduction of new combinations of the parental properties through merging the parent's genes which is done through the crossover with the connection specific distance coefficients. This increases the rate of the learning process by inheriting the knowledge from parents, which is termed as weight transmission. Pujol and Poli evolved weight, topology and activation functions of ANNs through genetic programming (Pujol and Poli 1997). Since pole balancing is a standard issue in the design of control systems, they tested the system for the development of a neural controller for a pole balancing problem; and obtained promising results.

Krishnan presented a method which could evolve the rules for changing the network weights, instead of the weights itself (Krishnan and Ciesielski 1994). Krishnan used an indirect encoding scheme where the gene represented a rule for changing the weights. They also applied the mutation and crossover operation of a standard genetic algorithm to genes until they obtained the desired weight adjustment function. This network was called the 2-Delta GANN (Whitley and Hanson 1989). This network performed better than the back propagation technique for the benchmark problems. For smaller problems, the back propagation technique was more effective, however according to the author; 2-Delta GANN was effective in solving problems which were known to be very difficult for BP. This technique also provided better results than other GANN which directly encoded the neural network weights in the chromosomes.

Yao explored all combinations of ANN parameters including: connections weights, architectures, learning rules and input features (Yao 1999). Yao explored evolving the neural architecture and found that evolution can find a near optimal ANN architecture automatically. Yao also evaluated direct and indirect genetic encoding scheme, concluding that direct encoding scheme is good at fine tuning and generating a compact architecture, while the indirect genetic encoding is superior for finding a particular type of ANN architecture quickly. He also explored various combinations of ANN parameter for evolution and concluded that evolving both ANN architectures and connection weights can produce better results.

Stanley presented a new type of TWEANN, the Neuro-Evolution of Augmenting Techniques (NEAT) (Stanley and Miikkulainen 2002). He identified three major challenges of TWEANNs and introduced solutions for them. His solutions include: "tracking genes with historical markings for easy crossover between different topologies", "innovative protection through speciation", and "starting from a minimal structure and making it complex with the passage of the generations". NEAT performed faster than many other neuro-evolutionary techniques. The complexity of NEAT network continue to grow during evolution. It starts with a very simple structure with no hidden neurons, and a simple feed-forward network of input and output neurons. During the course of evolution, the network continues to grow by addition of neurons to existing connections or by addition of a new connection between the unconnected neurons. NEAT doesn't involve the development of the neural network during the particular generation of evolution. It only updates its architecture from generation to generation. That is why NEAT is not a developmental model. The indirect method of NE is called the neural development, which will be discussed in the next section.

(Khan et al. 2013d; Khan and Zafari 2016) used CGP to introduce four different ways of evolving neural networks: Feed-forward CGP evolved ANN (FCGPANN) (Khan et al. 2013a), Recurrent CGP evolved ANN (RCGPANN) (Khan and Zafari 2016), Plastic CGP evolved ANN (PCGPANN) (Khan et al. 2013b), and Plastic Recurrent CGPANN (PRCGPANN).

In the first case, CGP is transformed to a feed-forward neural network by considering each node as a neuron, and providing each connection with a weight. The neurons of such a network are arranged in Cartesian format with rows and columns inspired by original CGP architecture, and later on restricted to a single row mostly giving the network an ability to create infinite graphs/topologies. Each neuron in the network can acquire connection from either a previous neuron or from the system input. Not all neurons are necessarily connected with each other or with system inputs, this provides the network with an ability to continuously evolve its complexity along with the weights. All the network parameters are represented by a string of numbers called genotype. The number of active neurons (connected from inputs to outputs), varies from generation to generation subject to the genotype selection. Output of any neuron or a system input can be a candidate for the system's output selection. The ultimate system functionality is identified by interconnecting neurons from output to input. FCGPANN was initially tested for its speed of learning, and evaluated against the previously introduced neuro-evolutionary techniques on benchmarks such as single and double Pole balancing (Khan et al. 2013d) showing superior performance in comparison to the previously introduced neuro-evolutionary techniques. FCGPANN is explored in a range of applications including: breast cancer detection, prediction of foreign currency exchange rates, load forecasting, internet multimedia traffic management, cloud resource estimation, solar irradiance prediction, wind power forecasting and arrhythmia detection (Nayab et al. 2013; Khan et al. 2013a, c; Arbab et al. 2014; Rehman et al. 2014a; Khan et al. 2014). FCGPANN outperformed all the previously introduced techniques as highlighted in the literature. The second type of CGPANN is the Recurrent CGPANN (RCGPANN). These networks are more suitable for modelling systems that are dynamic and nonlinear. This network is a modification to one of the earliest networks, the Jordan's network (Jordan 1986). In the Jordan's network there are state inputs that are equal in number to the outputs. These inputs are fed by the outputs through unit weights. The state inputs are present only at the input layer. In RCGPANN unlike the Jordan's network the state inputs can be connected, not necessarily to the first layer but to any layer. RCGPANN was also tested initially for its speed of learning similar to FCGPANN on both single and double pole balancing for both Markovian and non-Markovian cases. Its performance relative to other neuro-evolutionary techniques was superior. RCGPANN has been successfully applied to a number of applications including: Load forecasting, foreign currency exchange rates, bandwidth management and estimation (Khan and Zafari 2016; Khan et al. 2013c; Rehman et al. 2014b; Khan et al. 2013a) performing better than the previous neuro-evolutionary techniques.

Plasticity in neural networks has been the characteristic of choice when it comes to applications in dynamic systems due to its comparatively better performance (Papadrakakis et al. 1996; Sadeghi 2000; Carpenter and Grossberg 1988). The improved performance in Plastic neural networks can be attributed to the adaptability of its morphology to environmental stimuli. This is similar to the natural neural system. Plastic CGPANN has also been successfully applied to evolve a dynamic and robust computational model for efficiently predicting daily foreign currency exchange rates in advance based on past data (Khan et al. 2013b).

Plastic Recurrent Cartesian Genetic Programming Evolved Artificial Neural Network (PRCGPANN) is an online learning approach that incorporates developmental plasticity in Recurrent Neural Networks. Recurrent Neural Networks can compute random sequences of inputs due to their capability to acquire internal memory access. In a Plastic RCGPANN the output gene not only forms the system output but also plays a role in the developmental decision.

The research in artificial neural development is discussed in the next section.

4 Neural Development

The motivation behind the artificial neural networks was to replicate the computational models of the nervous system. ANN models mostly overlook the aspect that neurons present in the nervous system are part of the phenotype originated from the genotype through developmental procedure. Most of the aspects of the nervous system are determined from the information specified in the genotype (Kumar 2003). The genotype lays down the regulations for the development of the nervous system. The natural organisms have both the nervous system and genetic information stored in the nucleus of their cells (genotype).

The motive behind the development schemes is to increase the scalability of the ANNs, which is possible by having a minimum number of genes that can define the properties of the network instead of having a one to one relationship between the phenotype and genotype. These gene groups can influence several unrelated phenotypic traits with no dependency of the genotypic dimension on the phenotypic size. For example, there is a common estimation of 30–40 thousand genes in the human genotype (45 million DNA bases out of a total 10^9) while a mature phenotype consists of 10^{14} cells (Elliot and Elliot 2001; Lodish et al. 2003).

According to Parisi and Nolfi, the neural networks should be considered along with the genotypes to be viewed in biological context, as part of a population and inherited by the offspring from parents (Parisi 1997; Parisi and Nolfi 2001). Parisi and Nolfi utilized a growing encoding scheme (Nolfi et al. 1994; Nolfi and Parisi 1995) for evolving the architecture and the connection strength of the neural networks for controlling a small mobile robot (for a similar method see (Husbands et al. 1994)). The network comprises of a 2-D space having a group of the artificial neurons distributed with growing and branching axons. The genetic code provides the instruction for growth of the axons and the branching of neurons.

A neural development model, which starts with the single cell that undergoes the process of cell division and migration, was proposed by Cangelosi (Cangelosi et al. 1994). Every cell produces two daughter cells where the new cells are separated in a 2-dimensional space. This process of cell division and migration continues until a group of neurons which are arranged in a 2D space is produced. Finally, the neurons grow their axons to produce connections among each other. This process keeps going on until a neural network is developed. The rules for the cell division and migration are present in the genotype (for a related approach see (Dalaert and Beer 1994)).

Gruau also proposed a similar method (Gruau 1994). A single cell goes through various stages of cell division and differentiation until the development of a complete neural network. Every cell is divided into two daughter cells. The old connections are strengthened along with establishing new connections. The rules related to cell division and transformation lie in the genotype. The genotype of Gruau's model is similar to the binary tree structure of GP (Koza 1992). The top node of the genotype tree is the initial cell. Every node of the genotype in the tree encodes the operation of that cell, while the sub trees specify the operation which should be applied to the two daughter cells. As a result of following the tree using instructions in these cells, the neural network is developed.

As a result of further work done by Gruau, a method which was based on the genotype-phenotype mapping that allows the repetition of phenotypical structure by re-using the same genetic information was introduced. In this case, the terminal cell or nodes point to the other trees. This encoding method can result in complex phenotypical networks from compact genotype. Gruau termed this method as an "automatic definition of neural sub-networks (ADNS)" (Gruau 1994).

For evolving the parameters which grow into artificial neurons with bio-inspired morphology, Rust and Adams used a developmental model combined with a genetic algorithm. Although Rust and Adams were able to produce morphologies of neurons, they did not apply it to substantive problems (Rust et al. 2000; Rust and Adams 1999).

For dynamic neural growth mechanism in cognitive development, Quartz and Sejnowski provided a powerful manifesto (Quartz and Sejnowski 1997). Marcus also laid emphasis on the importance of the growing neural structures by using a developmental approach. In his words "I want to build a neural networks that grow, networks that show a good degree of self-organization even in the absence of experience" (Marcus 2001).

Jakobi developed an impressive artificial genomic regulatory network where the genes coded for proteins and the proteins either activated or suppressed the genes (Jakobi 1995). Jakobi defined the neurons, which had excitatory or inhibitory dendrites, through proteins. The individual cells divided and moved due to the interaction of the protein with the artificial genome. This resulted in the development of a multicellular system. After differentiation, every cell grew dendrites following the chemical sensitive development cones in-order to connect the cells. This resulted in a recurrent ANN capable of controlling a simulated Khepera robot for avoiding obstacles and navigation through corridors. The genotypes of every generation developed phenotypical structures, which were tested and the best one were chosen for breeding. Artificial evolutionary operations like cross over and mutation are utilized for creating offspring genotypes.

Various researches have studied the potential of Lindenmeyer Systems in developing ANNs and generative design (Lindenmeyer 1968). Boers and Kuiper adapted the L-systems for developing the architecture of the artificial neural networks (ANNs) i.e. a number of neurons and their interconnections (Boers and Kuiper 1992). A feed forward neural network was generated by evolving the rules of an L-System. They came to the conclusion that this methodology resulted in more modular neural networks, that performed better than the networks with the pre-defined structure.

Federici came up with an implicit encoding procedure for the development of the neuro-controller (Federici 2005). He also compared it with the direct scheme. He used adaptive rules relied on correlation between the post synaptic electrical activity and the local concentration of the synaptic activity and refractory chemicals.

Federici produced the neuro-controllers through two steps:

- He used a growth program in the genotype for developing the whole multi-cellular network in the form of the phenotype. This growth program inside every cell relies on local variables and is implemented by a simple recursive neural network which has a hidden layer (Similar to our use of CGP).
- During the second step, all the cells are translated into spiking neurons.

Every cell of the mature phenotype is a neuron of a spiking neuro-controller. The internal dynamics and synaptic properties of the corresponding neuron are specified by the type and metabolic concentrations of the cell. The topological properties of neurons such as its connections to the inputs, outputs and other neurons are produced by the position of the cell within organisms.

This network was implemented on a Khepera robot and the performance was tested both with direct and indirect coding schemes. Although the indirect method reached the high fitness faster, it had trouble in refining the final fitness value. Downing is in favour of a higher abstraction level in the neural development, because it avoids the complexities related with the axonal and dendritic growth. It also maintains the key aspects of the cell signalling, competition and cooperation of neural topologies in nature (Downing 2007). Downing also developed a system which he tested on a simple movement control problem called starfish. The task of the k-limbed animate is to move as far as possible in a limited time from the starting point. This produced positive preliminary results.

The next section explains one of the major problems with ANNs known as the 'catastrophic forgetting'.

5 Catastrophic Forgetting

Catastrophic forgetting is one of the main issues with ANN. During catastrophic forgetting, the network forgets the previous task; once it is trained to do a new task. The short term memory in The Human brain can be regarded as a forgetting problem with the Biological brain, but evolution has minimized that over time (Seipone and Bullinaria 2005). The problem is more catastrophic in traditional ANNs and it is a serious limitation in such models (McCloskey and Cohen 1989; French 1999). There are many methods available for either reducing or eliminating this problem. One of the basic reasons behind the catastrophic forgetting is interference in the shared weights (McCloskey and Cohen 1989; Ratcliff 1990). There are many methods used for reducing this interference such as sharpening algorithm for reducing the hidden unit activation overlap (connection usage) and the HARM model (Sharkey and Sharkey 1995); that implements a lookup table and divides the main task into two sub-tasks (French 1999; Seipone and Bullinaria 2005). There are also certain methods which use dual additive weights where the fast weights learn new tasks and slow weights are used for long term (Hinton and Plaut 1987). A large number of the methods rely on dual model architectures which consist of two distinct networks for processing early and long term storage processing (French 1991). The inspiration behind these methods is that human brains do not suffer from catastrophic forgetting as their brains evolve two different areas i.e. hippocampal system for learning the new information, and neocortical system for slow and long term learning and problem solving.

Brain has the capability of retaining information; some of this information might degrade over time in a gradual manner. Connectionist networks which are trained with a particular set of patterns when presented with new input patterns with no correlation to the old pattern, they adapt to the new patterns and completely forget the previous patterns. Robert addresses the problem of catastrophic forgetting in connectionist networks; it's consequences by highlighting the possible reasons that cause this behaviour and possible solution to this problem. According to Robert (French 1994), the problem of catastrophic forgetting can be alleviated by having separate areas for information handling and processing; and for retention of processed information.

"Conservative Training" and "Support Vector Reversal (SVR)" are presented in (Albesano et al. 2006) as solutions to mitigate the effect of catastrophic forgetting in ANNs in the domain of automatic speech recognition. In conservative training instead of assigning a value of zero to the missing units, target uses the output of the original network as an objective. While in SVR, support vectors are used to define the borders of the classes to keep the classification boundaries of the new network close to that of the originally trained network.

An Algorithm, elastic weight consolidation (EWC) inspired from the neurobiological model of synaptic consolidation; that is the mammalian brain is able to retain information as the excitatory synapses are strengthened. Continual learning is enabled by implementation of EWC which prevents the information of the previous task from being erased by reducing the plasticity of parameters from the previously learned task (Kirkpatric et al. 2017). Catastrophic interference can be seen in conjunction with a general dilemma coined by Grossberg (Grossberg 1980, 1982); the stability-plasticity dilemma which he published in his book in 1980. He states: "How can an organism's adaptive mechanisms be stable enough to resist environmental fluctuations which do not alter its behavioural success, but plastic enough to rapidly change in response to environmental demands that do alter its behavioural success".

Age limited learning effects are explained in the context of catastrophic forgetting by exploring the plasticity-stability dilemma in ANNs. In a parallel-distributed system, plasticity is essential for acquiring and incorporation of new information. Stability on the other hand is required to retain previously acquired knowledge. ANNs exhibit plasticity by readily adapting and learning new information at the cost of previously acquired knowledge (Mermillod et al. 2013). Human memory is emulated within a back propagation network by introducing grace degradation of information with the help of interleaved learning. Sparse encoding and activation function adjustment were also tested to assuage catastrophic interference. The results however revealed that they influenced the performance of the network but could not eradicate catastrophic forgetting in the network (Abdallah El Ali et al. 2008).

Robins' pseudo-rehearsal solution and French's activation sharpening algorithm were tested to overcome the problem of catastrophic interference with the former producing promising results. The solutions serve to reduce the catastrophic forgetting to some extent but fail a general solution to the problem (Ole-Marius et al. 2005). In Reinforcement learning (RL) problems, catastrophic forgetting can be prevented by avoiding overtraining and reasonably orthogonalising the input layer. To completely eliminate catastrophic forgetting into an RL agent, pseudo-rehearsal, a powerful continual learner can be adapted. Although, CHILD is a faster and more capable continual learner but due to its complex nature is difficult to execute (Cahill 2010).

These methods have reduced the catastrophic forgetting slightly; still the current models of ANNs cannot eliminate these problems. Although the ANN models have slightly adapted the biological neural structure, still they are not as complex as the biological neural systems. The biological neural systems can develop their own memory due to the changes in the synaptic connections, neural architecture, neurite growth, shrinkage and the variations of the chemical concentrations.

6 Conclusion

This chapter described various artificial neural networks, learning methods and their applications. Historical perspectives of evolutionary methods applied to ANNs were also elaborated. The chapter also presented a review of different methods for development of ANNs. The use of ANNs has spread to engineering and medical diagnostics. ANNs are the bio-inspired models of the brain. They have adopted some properties of biological neurons, but they are yet to match the complexity of the biological neurons.

The ANN models can perform efficiently in fixed task environment, however they seem to struggle with dynamically changing environment. As the learnt information in an ANN is encoded in the weights, retraining will cause the weights to change. This will affect the performance on the previous task. The performance of the network can also be affected; if the environmental conditions slightly change while the same task is being solved. Our network is yet to be tested on different task environments; however the weights and morphology of the network continue to develop during the task environment.

Our implemented system is inspired by the neuro-science. It also produced an artificial environment for the neurons. Our basic neuron model is based on biological study of neuron, their development and their mechanism of signal processing. The neurons can either grow more neurons or can die. They are able to produce complex neural structures based on the task requirement. We also evolved the rules for the model discussed in this book's development on the basis of neuro development techniques which were described earlier. Chalup proposed that an incremental scheme results in the development of the network in its stage of learning which would function more effectively than the artificially imposed inflexible system architecture (Chalup 2001). This argument supports the approach adopted by us.

The motivation for the model discussed is the work done on neuro-development techniques discussed in this chapter. The book evolves the rules for development of the neural architecture and their internal processing. It is evaluated on two learning environments i.e. the Wumpus world and the checkers, details are provided in Chaps. 6 and 7.

The next chapter will provide an insight into the design of the model along with biological inspiration in detail.

Chapter 5
Structure and Operation of Cartesian Genetic Programming Developmental Network (CGPDN) Model

This chapter presents detailed description of the CGPDN model.

- Key features and the biological basis of the CGPDN model
- General characteristics of CGPDN model
- Detailed description of neurons used in the CGPDN
- Information processing in the network
- Summary and concluding remarks.

The CGPDN model is very much influenced by the biological morphology of neurons and their arrangements. The neuron model deliberated here consists of a soma, dendrites, dendrite branches and axon branches. These neurons are arranged in a Euclidean grid based space. This gives a sense of virtual proximity and dynamic morphology to the branches. These neurons make synapses to the neighbouring branches; hence establishing communication among the neurons. The architecture of neurons is allowed to develop and evolve the genetic code inside neurons in search of desired intelligent behaviour.

1 Fundamental Attributes and Biological Basis for the CGPDN Model

In this section we present the main inspiration from literature and the key features incorporated into the CGPDN model. The section also explains the biological grounds of these concepts. All the characteristics of biological models integrated into CGPDN are enlisted in Table 1. The table also shows the presence and absence of these properties in existing ANNs and the neural development models.

Similar to the ANN models discussed earlier on, the system discussed in this book also consists of interconnected neurons. More bio-inspired features are incorporated

G.M. Khan, *Evolution of Artificial Neural Development*, Studies in Computational Intelligence 725, https://doi.org/10.1007/978-3-319-67466-7_5

Table 1 List of all the properties of biological systems that are incorporated into CGPDN or are present in ANNs and neural development models

S.No.	Name	ANNs	Neural development	Compartmental models	Biology	CGPDN
1	Neuron structure	Node with connections	Node with connections dendrites	Soma with dendrites, axon and dendrite branches	Soma with dendrites, axon and dendrite branches	Soma with dendrites, axon and dendrite branches
2	Interaction of branches	No	No	No	Yes	Yes
3	Neural function	Yes	Yes	Yes	Yes	Yes
4	*Resistance*	No	No	Yes	Yes	Yes
5	*Health*	No	No	No	Yes	Yes
6	*Weight*	Yes	Yes	Yes	Yes	Yes
7	Neural activity	No	No	No	Yes	Yes
8	Synaptic communication	No	No	Yes	Yes	Yes
9	Arrangement of Neurons	Fixed	Fixed	Fixed	Arranged in space (Dynamic Morphology)	Arranged in Artificial space (Dynamic Morphology)
10	Electrical signalling	Yes	Yes	Yes	Yes	Yes
11	Chemical signalling	No	No	No	Yes	No
12	Developmental plasticity	Yes	No	No	Yes	Yes
13	Spiking (Information processing)	Yes, but not all	Yes, but not all	Yes, but not all	Yes	Yes
14	Arbitrary I/O	No	No	No	Yes	Yes
15	Learning rule	Specified	Specified	Specified	Unspecified	Unspecified

into the neurons of this system and have close resemblance to neuro-developmental models. These bio-inspired features are the morphological features. Every neuron contains different number of dendrites with each dendrite having a number of branches for the reception of the input signals. Axon being the major compartment of the neuron is responsible for output of the system. The axon comprises of a number of branches. The dendrite and axon branches are provided with the flexibility to grow and shrink and produce offspring (new dendrite and axon branches). The system is capable of generating its own structure based on the functionality required for the desired problem. The neurons of the model discussed in this book have three key morphological components.

- The dendrite with branches which can receive and process inputs.
- Cell body which can process signals received from the dendrites.
- An axon which can transfer signals to other neurons through axon branches.

Traditional ANNs only have nodes and connections, they don't have any concept of branches or axon and dendrites. The neural development methods discussed in Chap. 4 Sect. 4.5 introduced the concept of axon and dendrite branches, which develop into a network.

The architecture of both ANNs and neural developmental systems remain fixed during the evaluation period, they are changed only during the evolutionary period. The biological neurons exist in space. They interact with each other and move their branches from one place to another. The model of the book also adapted similar mechanism which places neurons with its dendrite and axon branches in Euclidean grid space. This places them in close proximity to each other so that they can interact. The axon and dendrite branches can navigate in the space, causing the morphology of the network to change; while solving the problem. Though the ANNs and neural development sometimes consider the connections between the neurons as dendritic; still they are yet to replicate the biological dendrites in the types of morphology that exist. The signals of different dendritic branches can interact with each other in biological dendrites, while such interaction do not exist in ANNs and neural development. The model of this book has adopted this feature and evolved the functions for the interaction between the branches, since there is no precise mathematical model for approximating these functions.

The neuron branches have the property of resistance which depends upon branch length and affects the signals propagating through them. However, there is no concept of branch lengths in ANN literature or neural development models. The model discussed in this book adapted this property for biological axon and dendrite branches. Resistance affects the signals which propagate in these branches.

The biological neurons have the property of 'health' which affects the signal processing inside neurons and neurites. This property is also not present in the previously introduced ANN and neural-development models. The incorporation of the health property in the model allows the neuron or neurites to replicate and die. The synapses can transfer electrical signals between the neurons. The ANN and neural development literature consider the synapses as only the contact point; however the biological synapses provide a complex mechanism for signal transfer and modulation. The CGP programs have been evolved for finding out the useful mechanisms which allow signal transfer across a synapse. In biological brain, spikes are responsible for signalling; which is also used in some of the ANN and neural development models. The model in this book has also used a similar mechanism, as the signals are transferred to other neurons only if the neuron fires. The potential of the branches is affected by the synapses in biological systems through the changes in the concentration of chemicals (ions in the space between the neurons). Though the synapses in the

CGPDN are not chemical, the synapse made in the CGPDN updates the weights and potential values of the neighbouring branches. It is similar to the chemical changes at the synapses.

The biological neurons are constantly changing. Their internal processes and morphology change all the time in response to the environmental signals. The external environmental signals affect the development process of the brain, which is termed as "developmental plasticity". It usually occurs in the form of synaptic pruning (Van Ooyen and Pelt 1994).

The developmental plasticity eliminates the weaker synaptic contact, but it preserves and strengthens the stronger connections. The decision about keeping or pruning the connections is made on the basis of the common experiences which generate similar sensory inputs. The frequently activated connections are preserved. The process of apoptosis results in the death of the neuron. The neurons are damaged and then they die. The plasticity assists the brain in adapting to the environment.

The model discussed in this book also incorporates a form of developmental plasticity due to which branches can prune as well as new branches can grow. The "life cycle" chromosome controls this process. This chromosome determines whether the branches are to be pruned or new branches are to be introduced. Every time a branch is active, a life cycle program runs to determine whether the branch should be removed or it should continue to be a part of the processing. The life cycle also determines if there is need for any daughter branch in the network.

Starting from the randomly connected network, the new branches are allowed to navigate (they can move from one grid sequence to another resulting in new connections) in the environment on the basis of the evolutionary rules explained in Chap. 5, Sect. 4. An arbitrary connectivity sequence can alleviate the time spent in discovering the connection in the early developmental stage of neuron. Most of the neural development techniques which were described earlier (See Chap. 4, Sect. 4) start with a single cell which then develops into a complete network before they can be processed.

Plasticity is a very important concept for understanding the neural systems. It is the ability of permanently changing or reforming which manifests at various levels in the nervous system. Synaptic plasticity is responsible for the dynamic abilities of the neural system. The synaptic plasticity represents the variation in the synaptic transmission (Debanne et al. 2003). The synaptic plasticity can occur both at the post-synaptic and the pre-synaptic levels. It might involve the variation in the post-synaptic excitability (the probability of generating action potential in response to the fixed stimulus), which depends on the previous pattern of the input (Gaiarsa et al. 2002). Synaptic activity can also alter the number of receptors (sites of the neurotransmitter action) on the membrane (Frey and Morris 1997). These procedures can communicate, which results in positive feedback effects. Certain cells might never fire while others might saturate at some maximal firing rate.

Synaptic plasticity is integrated in the CGPDN through the introduction of three types of weights one for each:

(1) Dendrite Branch
(2) Soma, and
(3) Axon Branch

During the development of the network, the genetic processes are used to adjust the weights. The changes in the weights of dendrite branches are analogous to the amplifications of a signal along the dendritic branch (see (London and Husser 2005)), while the changes in axon branch weight are analogous to the variations in the pre-synaptic level and post-synaptic level (at synapse). The observation that a fixed stimulus induces myriad responses in a variety of neurons, justifies the inclusion of soma weight (Traub 1977).

The synaptic plasticity in the SNN models is primarily based on the work of Heb, which is based on the principal that repeated and persistent stimulation of a post-synaptic neuron by a pre-synaptic neuron results in the strengthening of the connection between the two cells. The recent development in this idea is to use the spike time dependent plasticity (STDP) rules in order to update the weights in SNN network (Roberts and Bell 2002; Song et al. 2000).

The variations in the *weights* are dependent upon the relative timing of the pre-synaptic and post-synaptic spikes. Various neural systems have documented this process. There have been numerous rules proposed in which *weights* either increase, decrease or remain unchanged, reviewed in (Roberts and Bell 2002). These schemes result in interesting patterns of global behaviour, including competitions between synapses, despite the fact that these schemes use an update rule; which depends on the firing time of only two neurons (Van Rossum et al. 2000).

Some aspects of the above ideas are adopted in the CGPDN system by allowing the three different types of *weights* that vary in response to the firing of neurons. With every fire of neuron, the life cycle and weight processing chromosomes run; which updates their values (details present in the CGP neuron sections). The variations in *weight* allow neurons and branches to become more active when they are involved in processing data. The next section will explain that the active dendrite branches, axon branches and soma are updated more often as their genetic code is executed as long as they are operational. This interaction influences the developmental and evolutionary processes by the sequence of activity transmitting through the system. This interaction is considered to be significant for the biological neural systems, in which the sensory input and other environmental parameters are of importance in defining evolutionary facets (Kandel et al. 2000).

Plasticity is an emergent property of the brain for developing memory, while it is mostly imposed on ANN models. The model presented in this book, adapts plasticity by allowing the network to develop its own plasticity; rather than imposing it. The CGPDN model idealizes the behaviour of neuron in terms of seven main processes. The first three (3) are electrical processes, while the next three (3) are life cycle (developmental) mechanism and the last one (1) is weight processing (chemical concentration):

(1) Electrical Processing in dendrite: Local interaction among neighbouring branches of the same dendrite.
(2) Electrical Processing in Soma: Processing of signals which are received from dendrites at the soma, and the decision about firing an action potential.
(3) Electrical Processing in Axo-Synaptic branch: Synaptic connections which transfer potential through the axon branches to the neighbouring dendrite branches.
(4) Life Cycle of Dendrite branch: The growth and shrinkage of the dendrite branch as well as production of new dendrite branches, and removal of old branches.
(5) Life Cycle of Axo-Synaptic Branch: The growth and shrinkage of the axon branch as well as the production of new axon branches along with the removal of old branches.
(6) Life Cycle of Soma: Creation or destruction of neurons.
(7) Weight Processing: Updating the synaptic weights (Consequently the ability to make synaptic connections) between the axon branches and neighbouring dendrite branches.

A separate chromosome (CGP program) is used to represent every aspect. The next section will describe the model of CGPDN, its sub-parts, its evolutionary strategy and its interfacing with the external environment.

2 The CGP Developmental Network (CGPDN)

This section explains the detailed structure of the CGPDN, the rules and the evolutionary strategy which is used for evolving the system. There are two main aspects associated with the CGPDN.

(a) A phenotype in the form of Neurons comprising of Axon and a number of Dendrites, with Axon and each Dendrite having a number of branches.
(b) A genotype representing the genetic code of the neurons. Every genotype has seven chromosomes where every chromosome is represented with a digital circuit.

Aspect (a) is mainly related to the neural components and their properties while aspect (b) deals with the internal behaviour of the neurons in the network. The functionality of different parts of the neuron is represented by chromosomes. During evolution, the second aspect (genotype) evolves towards the best functionality while the first aspect (neural components and their properties) only varies while it is performing the learning task, i.e. during its lifetime. The CGPDN has been organized in such a way that all the neurons are placed randomly in a two dimensional grid i.e. the CGPDN grid; that is why they only know about their neighbours (as shown in Fig. 1). The users specifies the initial number of neurons. Every neuron initially has a random number of dendrites, dendrite branches, an axon and a random number of axon branches. Neurons receive information through dendrite branches, and then transfer this information to the neighbouring neurons through axon branches. The

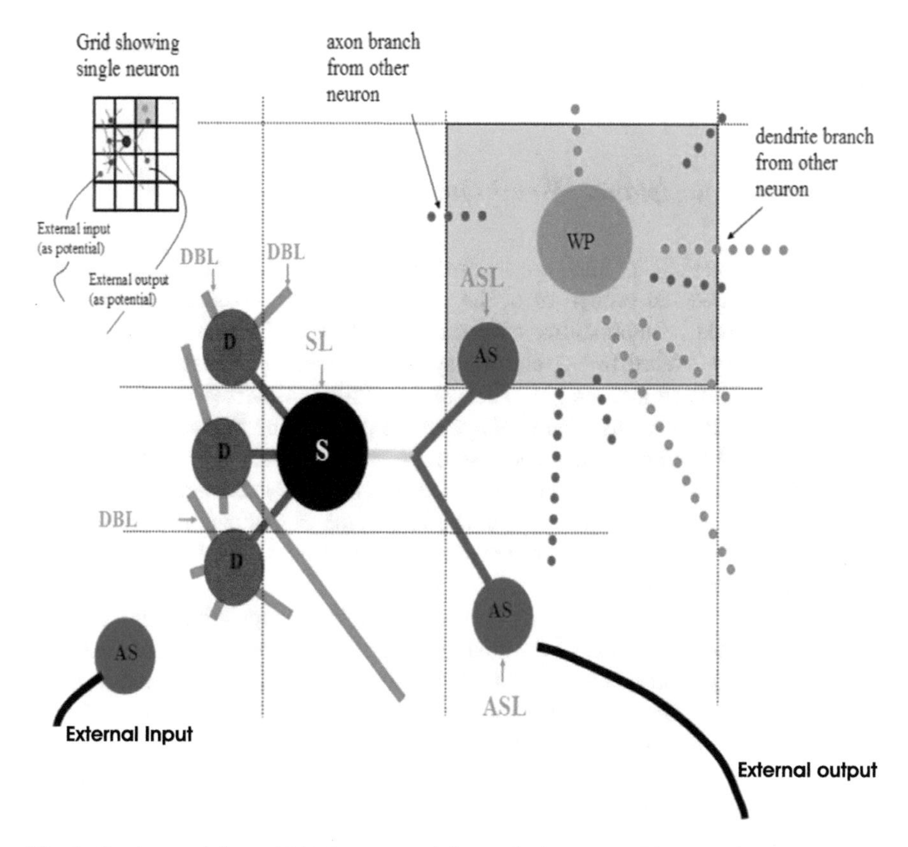

Fig. 1 On the top left a grid is shown containing a single neuron. The rest of the figure is an exploded view of the neuron. The neuron consists of seven evolved computational functions. Three are electrical and process a simulated potential in the dendrite (D), soma (S) and axo-synapse branch (AS). Three more are developmental in nature and are responsible for the life-cycle of neural components (shown in grey). They decide whether dendrite branches (DBL), soma (SL) and axo-synaptic branches (ASL) should die, change, or replicate. The remaining evolved computational function (WP) adjusts synaptic and dendritic weights and is used to decide the transfer of potential from a firing neuron (dashed line emanating from soma) to a neighbouring neuron

dynamics of the network vary during this process; the branches might grow or shrink and can move from one CGPDN grid point to another. They can also produce new branches and can even disappear. Neurons might die or even produce new neurons. The axon branches transfer information only to the nearby dendrite branches. This process takes place by passing signals from all the neighbouring branches through a CGP program. This acts as an electro-chemical synapse, and it updates the potential values only in neighbouring branches. The communication between the neurons and the internal processing of neurons takes place through the electrical potential which is represented by an integer. Before the inputs and outputs are applied to the network, they are also translated in terms of potentials (integers). The next four subsections explain the parameters of CGPDN (*Resistance*, *Health*, *Weight* and *Statefactor*),

Cartesian Genetic Program (used as genotype), Evolutionary Strategy, and the way inputs and outputs are applied to the network.

2.1 *Health, Resistance, Weight and Statefactor*

There are four variables which are incorporated into the CGPDN. They represent either the fundamental properties of the neurons (*health*, *resistance*, *weight*) and are used as an aid to computational efficiency (*statefactor*). There are three variables assigned to every dendrite branch and axo-synaptic connection. These three variables are *health*, *resistance* and *weight*. The values of these variables can be adjusted through the CGP programs. The *health* variable governs the replication and death of dendrites and axon branches. The *resistance* variable controls the growth and/or shrinkage of dendrites and axon branches. The *weight* variable is used for calculating the potentials in the network. Every soma contains the *weight* and *health* variables. Figure 9 summarizes the use of these variables. The *health*, *weight* and *resistance* are represented by integers.

Statefactor is a parameter used for the reduction of the computational burden. This is achieved by keeping some of the neurons and branches inactive for a number of life cycles. A 'zero' *statefactor* indicates that the neurons and the branches are active and their corresponding program is run. The CGP programs influence the value of *statefactor*, since it relies on the CGP electrical processing chromosomes (explained later). *Statefactor* is inspired from the characteristic of biological systems that not all neurons and dendrite branches are effectively participating in every process.

2.2 *Cartesian Genetic Program (Chromosome)*

Multiplexer are used as the CGP nodes function (Miller et al. 2000). Every function node has three inputs and an output. There are four set of functions used as shown in Fig. 2. The multiplexers can be thought of as atomic in nature as they can be used for representing any logic function (Chen and Hurst 1982; Miller et al. 2000).

The inputs are A, B and C in the Fig. 2. These functions can be either arithmetic (Refer to Chap. 6, Wumpus world) or Boolean (Refer to Chap. 7, Checkers). In case

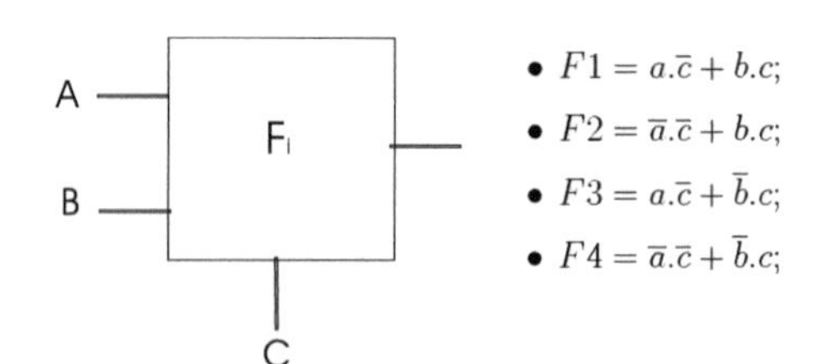

Fig. 2 Multiplexer diagram, showing inputs A, B and C, and function Fi. Figure also lists all the four possible functions that can be implemented by multiplexer

of arithmetic these operations are additions (+), multiplications (.) and complements. All the inputs are considered to be 8-bit integers. In case of Boolean functions, these operations are either logical AND (.) or logical OR (+). There are four genes required for the multiplexer to describe the type of multiplexer (underlined in Fig. 3) and its connections. The multiplexer operate bit-wise on 32-bit data in case of Boolean operations and 8-bit integers in case of arithmetic operations.

The model in the book initially uses arithmetic operations on 8-bit data as it performs slowly with high bit values. The discrepancy associated with using the arithmetic operations with fixed data is the loss of information during the multiplication process through the removal of the lower 8-bits. Boolean bitwise operations do not result in the loss of information, however all the bits are independent. The arithmetic operations require more processing time compared to the Boolean operations, which causes the reduction in the speed of evolution; since multiplication is involved causing increase in computation time.

The function set and other parameters can be modified, and new methods can be introduced to have linear/non-linear function sets to find the unknown functions in DNA of neuron. The method described in this book, pitch the idea, and the function sets are not the ultimate.

The genotype and the obtained corresponding phenotype are shown in Fig. 3 where they connect the nodes as specified in the genotype. Figure 3 also shows the inputs and outputs of the CGP. The output is taken from the nodes 6, 8 & 4 as specified in the genotype. The model of this book does not specify the output in the genotype and uses a fixed pseudo random list of numbers for specifying location of the output.

As mentioned earlier (Chap. 3 Sect. 3.3), there is only one row and number of columns are the same as the number of nodes. User defines the maximum number of nodes. These nodes are not necessarily all connected. There are two ways to apply the inputs to the CGP chromosomes. They are

- Scalar
- Vector

If the inputs are applied in a scalar way, then the inputs and outputs are in form of integers while in case of vector, the inputs which are required by the chromosomes; are arranged in the form of arrays. The array is then divided into 10 CGP input vectors. If the total number of inputs cannot be divided into ten equal parts then they are padded with zeroes. This allows the CGP circuit chromosome to process an arbitrary number of inputs by clocking through the vector's elements. Generally, the CGP cannot take variable number of inputs at run time. Since the inputs are arranged in vector form, where every vector has arbitrary number of elements; this method will result in some noise. The noise will be more pronounced if the number of inputs are less than ten, because we padded it with zeroes when the number of inputs could not have been divided into ten sub vectors. The increase in the number of inputs will cause the noise to decrease.

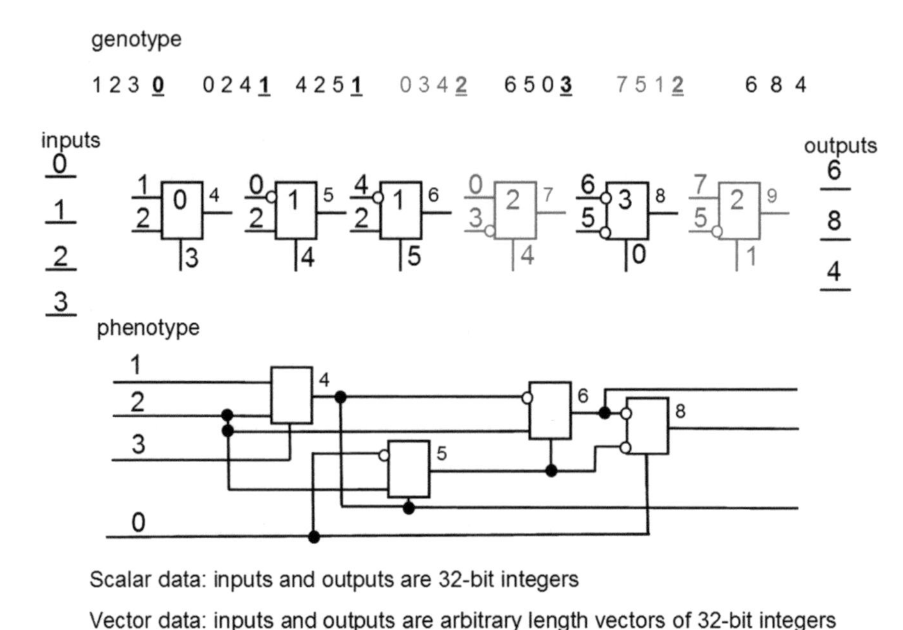

Fig. 3 Structure of CGP chromosome. Showing a genotype for a 4 input, 3 output function and its decoded phenotype. Inputs and outputs can be either simple integers or an array of integers. Note nodes and genes in grey are unused and small open circles on inputs indicate inversion. The function type in genotype is indicated by underline (underneath the integer showing function of multiplexer). All the inputs and outputs of multiplexers are labeled. Labels on the inputs of the multiplexer show where are they connected. Input to CGP is applied through the input lines as shown in figure. The number of inputs (four in this case) and outputs (three in this case) to the CGP are defined by the user, which is different from the number of inputs per node (three in this case i.e. a, b and c.)

2.3 Evolutionary Strategy

The evolutionary strategy used in this work is of the form 1+λ, where λ is set to 4 (Yu and Miller 2001). This means that it has one parent with 4 offspring so the population size is 5. The parent remains unvaried while the offspring are produced by the mutation of parents. The best chromosome gets promoted to the next generation. In case, if there is a tie between the highest fitness of two chromosomes, then the one genetically newer is selected (Miller et al. 2000).

The evolutionary cycle occurs in the following steps.

- A random population of 5 genotypes is created where each genotype consists of seven chromosomes of neurons.
- A CGPDN with a user defined initial number of Neurons with random number of dendrites and branch structures is created.
- An evolutionary generation consists of:

  - For every genotype C in the population, a random copy of the CGPDN is produced. Then the genotype is provided to the CGPDN, which is run on target application. The Fitness F(C) of the resulting CGPDN is calculated for the task scenario.

- Among the population, the one with the best fitness F(C) is selected. In case if there is a tie between the fitness, then the newer one of them is selected (Miller et al. 2000).
- A new population is created through mutation, while the promoted genotype remains unchanged. This process continues until either maximum number of generations or a solution is found.

Initially random network of neurons, dendrites, dendrite branches and axon branches can grow into a mature network by the execution of the program encoded in the genotype.

The promoted genotype is then mutated for the production of new genotypes (offspring) as under:

(1) The number of genes which are to be mutated, are calculated first. Mathematically,
Number of bits to be mutated= number of genes X mutation rate/100
Where:
The number of genes = (number of inputs per node (3 in this case) + 1) X number of nodes per chromosomes X number of chromosomes (7 in this case)
(2) The genes are selected pseudo-randomly one at a time, and they are also mutated pseudo-randomly. The mutation of gene means;

  - If it is a connection, it is replaced with another connection.
  - If it is a function, it is replaced with another function.

The evolutionary strategy 1+λ is used because it performs better for CGP, in case different genetic programming or machine learning is used for finding DNA function set of neuron, then a corresponding strategy best for those algorithms should be introduced. Also rank based selection is used in this book, although probabilistic selection might perform better, and can be explored in future.

2.4 Inputs and Outputs

The axo-synaptic electrical processing chromosomes can be used to apply inputs to the CGPDN through axon branches. Just like the axon branches of neurons shown in the Fig. 4, these branches are distributed in the network. The branches can be considered to be the "input neurons". The input is taken from the environment and then transferred directly to the input axo-synapses. The programs encoded in the axo-synaptic electrical branches are executed to apply input to the system. The resulting signal is then transferred to the neighbouring active dendrite branches.

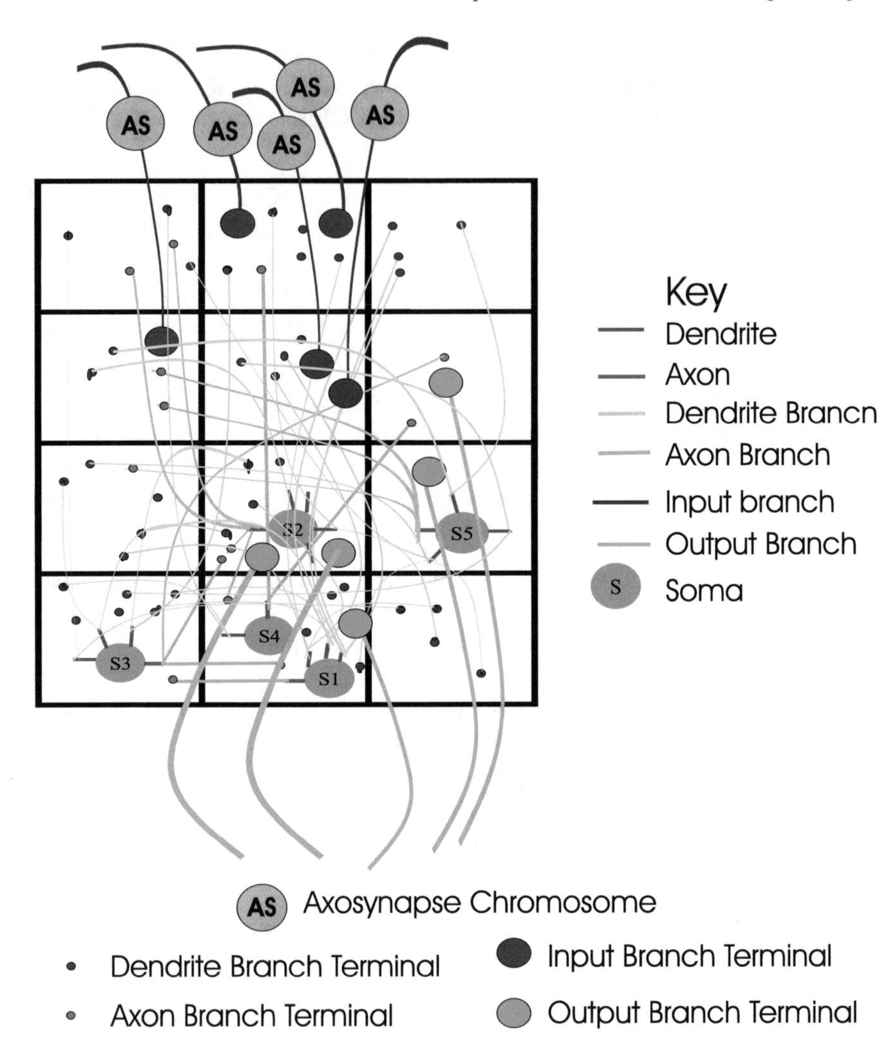

Fig. 4 A schematic illustration of a 3 × 4 CGPDN grid having five neurons, with each a number of dendrites comprises of branches, an axon with branches. Inputs are applied using input axo-synapse branches located 5 different locations by running axosynaptic CGP programs. Outputs are taken from five random locations through output dendrite branches

Similarly there are output neurons which can read the signals from the system through the dendrite branches. These output dendrite branches are distributed across the network as shown in Fig. 4. Just like other dendrite branches, that are part of system neurons, the axo-synaptic chromosomes of neurons update these branches as well. After every five cycles, without further processing, the output from these output dendrite branches representing output neurons is taken. The number of inputs and outputs can vary at run time (during development). An existing branch can be removed, or a new branch for input or output may be introduced into the network. Due to this CGPDN can handle arbitrary number of inputs or outputs at run time.

Next section will describe the complete neuron model along with its sub-processes.

3 CGP Model of Neuron (The Genotype)

The neural functionality of the model under deliberation is divided into three major categories:

- Electrical Processing
- Life Cycle
- Weight Processing

3.1 Electrical Processing

The electrical processing part is responsible for signal processing inside neurons and communication between the neurons. There are three chromosomes in the electrical processing part (as shown in Fig. 6).

- Electrical Processing in dendrite.
- Electrical Processing in Soma.
- Electrical Processing in axo-synaptic branch.

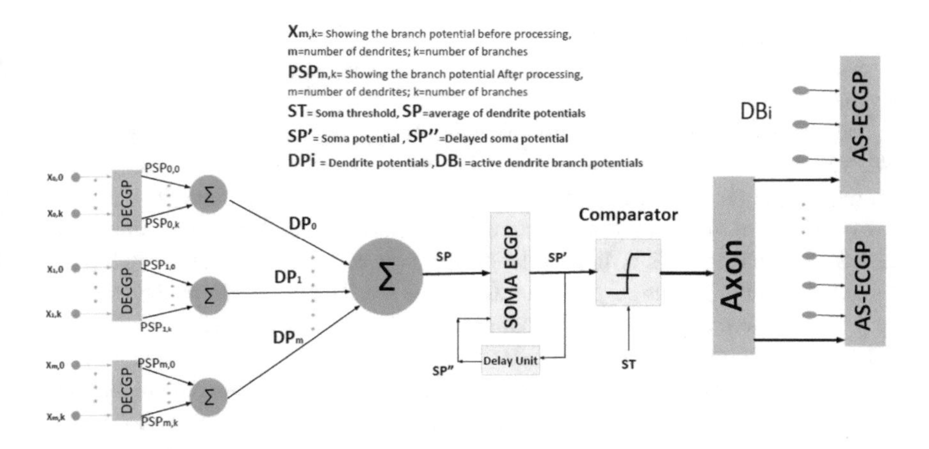

Fig. 5 Electrical processing in neuron at different stages, from left to right branch potentials are processed by DECGP, then averaged at each dendrite, and soma, which processes it further using the Soma-ECGP giving a final soma potential. This is fed in to a comparator which decides whether to fire an action potential. This is transferred using the AS-ECGP

The manner in which the electrical signals are processed and transferred to other neurons is depicted in Fig. 5.

3.1.1 Electrical Processing in Dendrite

It is a vector processing chromosome, which handles the interaction between the potentials in different dendrite branches belonging to the same dendrite. The inputs and outputs are shown in Fig. 6. The inputs are made up of the potentials of all the active branches, which are connected to the dendrite and the soma potential. Practically there are many dendrite branch potentials and one potential of soma. The importance of soma potential is made more prominent by increasing its entries (equal to the number of active dendrite branches) in the input vector before applying it to the CGP program encoded in the chromosome. The updated values the dendrite branch potentials are produced by the CGP program as the outputs. The input and outputs to the Dendrite Electrical Processing CGP (DECGP) can be seen in Fig. 6. The updated values of the branch potentials are further processed. The processing of the potential of each branch takes place through the addition of the weighted values of *Resistance*, *Health*, and *Weight* by utilizing the following equation:

$$P = (\acute{P} + \alpha_D H + \beta_D W - \gamma_D R)\&\ \text{Mask} \tag{1}$$

where;

P = Processed Potential

$\acute{P}$ = Updated potential respectively

H = Health of the dendrite branch

W = Weight of the dendrite branch

R = Resistance of dendrite branch

α_D, β_D and γ_D = adjustment parameters whose values are between 0 and 1 (usually defined by user).

Here in this case, its values are 0.02(2%), 0.05(5%) and 0.05(5%) respectively. We have used these values intuitively, but they can be evolved or looked deeply into biology to get the best optimum. The masking of the processed potential can avoid the overflow and it is based on the number of bits used for processing. Equation 1 also shows that there is a direct relationship between the *health* & *weight* of the branches and the potential. The increase in the *health* and *weight* of the branches will cause an increase in the potential as well. The direct relationship between the *health*, *weight* and potential is justified by that fact that healthy branches facilitate the flow of potential, while *weights* are responsible for the amplification of potential. The equation also indicates that there is an inverse relationship between the *resistance* and potential (which is a common behaviour of resistor). An increase in the *resistance* will result in the decreased potential.

The adjustment of *statefactor* of the branches takes place on the basis of the updated value of the branch potential. An increase in the activity of the branch takes place with the increase in the change in potential during the DECGP process and

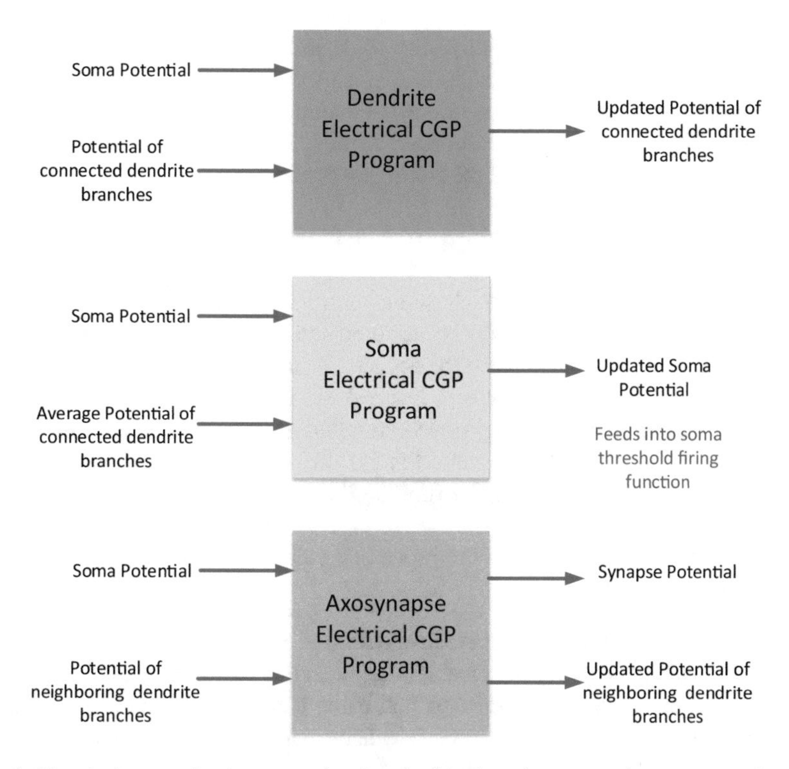

Fig. 6 Electrical processing in neuron showing dendrite branch, soma and axosynapse electrical CGP programs with their corresponding inputs and outputs

vice versa. The motive behind this is encouraging the sensitive branches to participate in the process by keeping them active. A range of thresholds are arranged for the *statefactor*. If any of the branches have its *statefactor* equal to zero i.e. they are active; and its life cycle CGP program is run. However if the branch is not active, the processing of other dendrites is initiated, and same is repeated for all the dendrites and their corresponding branches. After processing all dendrites, the average value of all the dendrite potentials is taken. It is also the average value of all the active dendrite branches attached to them. This potential along with the soma potential are applied to the CGP soma electrical processing chromosome as inputs. The details will be explained in the following subsections.

3.1.2 Electrical Processing in Soma

This is a scalar processing chromosome which is responsible for determining the final value of soma potential after all the signals are received from the dendrites. The dendrite potentials are averaged which are also the averages of potentials of active branches attached to them as shown in Fig. 5. This average potential along with the soma potential is applied as input to the soma electrical processing chromosome (Soma-ECGP) as shown in Fig. 6.

The chromosome also produces an updated value of the soma potential (($\acute{P}$)) as the output. It is then further processed with a weighted summation of *Health (H)* and *Weight (W)* of the soma by using the following equation.

$$P = (\acute{P} + \alpha_S H + \beta_S W)\&\ \text{Mask} \tag{2}$$

where the values of α_S and β_S are chosen to be 0.02 and 0.05.

The processed soma potential is then compared with the threshold potential of the soma. This comparison helps in the decision regarding the firing of the potential. In case the soma fires, then it is kept inactive for a few cycles (refractory period) by varying its *statefactor* and its threshold value is also adjusted to a new value (maximum). Following this, the soma life cycle chromosome is run and the firing potential is sent to other neurons by running the axo-synapse electrical processing chromosome.

If the soma does not fire then the processed potential value is checked and following actions are taken:

- If the soma potential value is less than one third of the maximum value, its statefactor is set to a higher value; such that it is kept inactive for three cycles (a parameter that can be changed). This indicates that if the potential of soma is a low value, which makes it unable to fire; then it is kept inactive for more time. The firing somas are encouraged and kept more active.
- If the soma potential value is more than one third but less than the half of the maximum value (a parameter that can be changed), then it is kept inactive for one cycle (a parameter that can be changed).
- If the soma potential is higher than half of the maximum value, then it is kept active for the next cycle and its life cycle program is run. This indicates that somas with higher potentials are kept more active.

3.1.3 Electrical Processing in Axo-Synaptic Branch

It is a vector processing chromosome in which the potential is transferred from soma to the other neurons through the axon branches. Both the axon branches and the synapse are thought to be a single entity with combined properties. The axo-synapses transfer the signal only to the neighbouring active dendrite branches as described in Fig. 7. Those branches which share the same grid square form a neighbourhood. Figure 6 shows both the inputs and outputs of the chromosome responsible for the electrical processing in each axo-synaptic branch. As discussed earlier, the potential of the soma is biased through the introduction of the multiple entries of potential of the soma for its increased impact.

This chromosome produces an updated value of the neighbouring dendrite branch potentials. It also yields the updated value of the axo-synaptic potential as output.

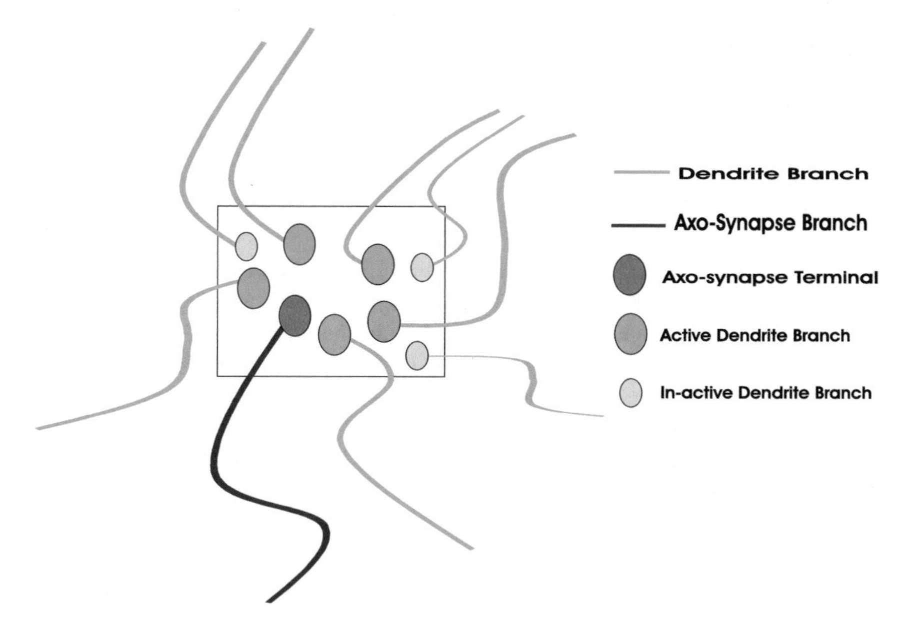

Fig. 7 Diagram showing one of the grid squares in which signal is transferred from axo-synapse to dendrite branches, showing inactive and active branches

Equation 3 is used for processing the axo-synaptic potential as the weighted summation of the *Health*, *Weight* and *Resistance* of the axon branch.

$$P = (\acute{P} + \alpha_{AS}H + \beta_{AS}W - \gamma_{AS}R)\&\ \text{Mask} \tag{3}$$

where;

P = Potential

$\acute{P}$ = Updated Potential

H = Health of axon branch

R = Resistance of the axon branch

W = Weight of the axon branch

α_{AS}, β_{AS} and γ_{AS} = The adjustment parameters which can have values between 0 and 1.

In this case, there values are 0.02, 0.05, and 0.05 respectively.

After the above process, the axo-synaptic branch weight processing program (shown in Fig. 8) runs and the processed axo-synaptic potential is assigned to the dendrite branch which has the highest updated Weight. The updated value of branch potential is used to adjust the *statefactor* of branches. There is a direct relation between the activeness of branch and the change in the potential during the Axo-Synapse Electrical CGP (AS-ECGP). The branch will become more active with the increase in the change in the potential during the AS-ECGP and vice versa. The life

CGP program of the active branches is run. The axo-synaptic branch CGP is run in all the active axon branches one by one.

3.2 Weight Processing

It is also a vector processing chromosome which updates the *weights* of the branches. It is just made up of one chromosome. The *weights* of axon and dendrite branches affect their ability of modulating and transferring the information efficiently. The modulation of potential is the responsibility of the weights. Weights affect almost every neural process either by being an input to a chromosomal program or by being a factor in the post processing of the signals.

Figure 8 shows the inputs and outputs of the axo-synaptic weight processing chromosome. The CGP program, encoded in the chromosome, takes the weight of an axo-synapse and the neighbouring dendrite branches as its input; while its output is their updated values. The synaptic potential produced at the AS-ECGP, is assigned to the dendrite branches which have the highest weight after weight processing.

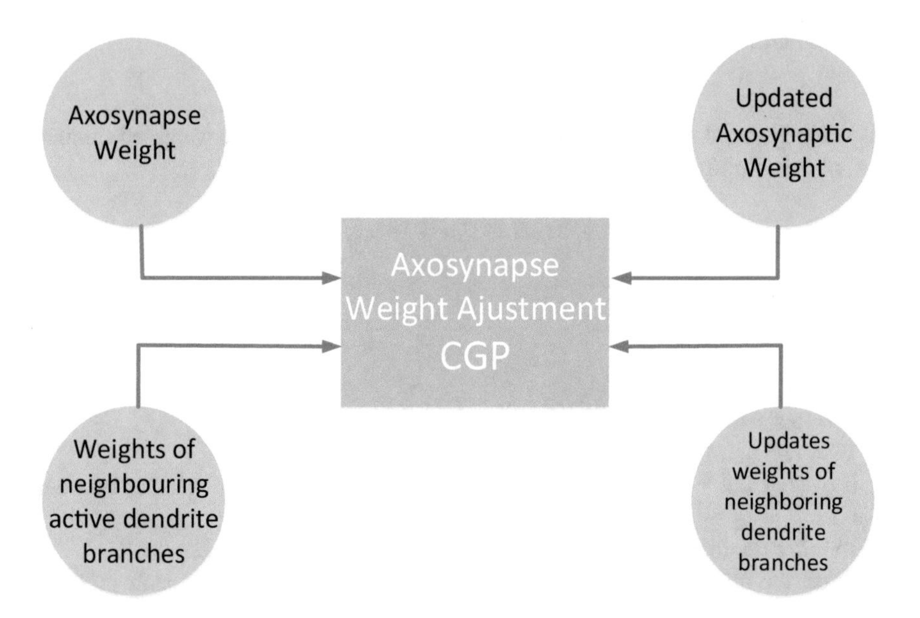

Fig. 8 Weight processing in axosynaptic branch with its corresponding inputs and outputs

3.3 *Life Cycle of Neuron*

The life cycle of neuron is responsible for either the growth or the decrease in the number of neurons and neurite branches. It is also responsible for the migration of the neurite branches. There are three chromosomes in the life cycle of neurons:

- Life Cycle of Dendrite branch
- Life Cycle of Soma
- Life Cycle of Axo-synapse branch

3.3.1 Life Cycle of Dendrite Branch

It is a scalar processing chromosome. Figure 9 shows the inputs and outputs of the chromosome. This process updates *resistance* and *health* of the branch. The variation

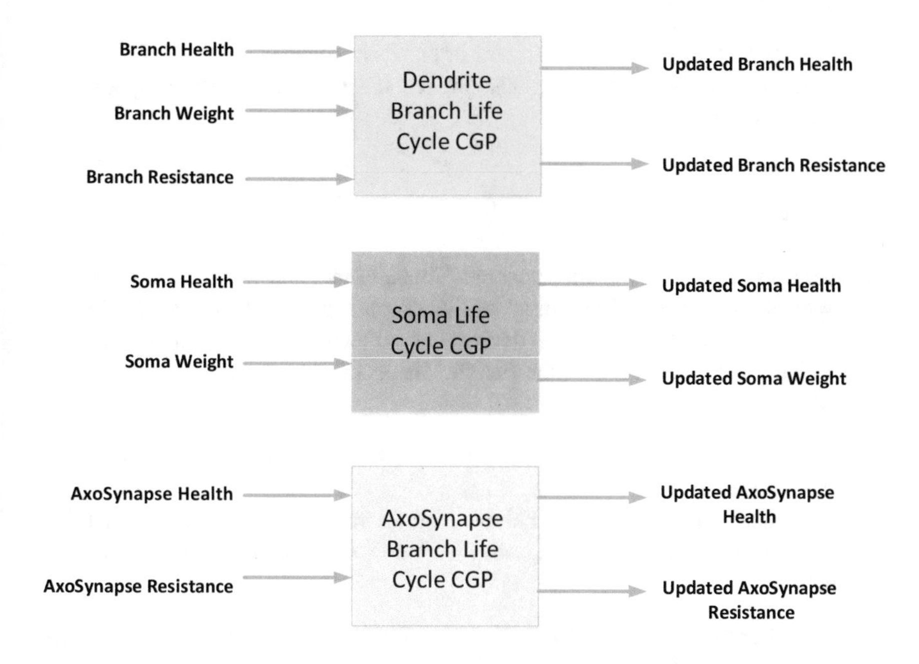

Fig. 9 Life cycle of neuron, showing CGP programs for life cycles in dendrite branch, soma, and axosynapse branch with their corresponding inputs and outputs

in *resistance* of the dendrite branches is utilized to decide whether it will grow, shrink or remain unchanged. The branch can migrate randomly to a different neighboring location if the resistance variation, during the process goes above the threshold (R_{DB}). There are 8 possible neighbouring squares in the rectangular grid, where the branch can migrate. However its movement is restricted to only one square at a time. The variation in resistance can be either negative or positive i.e. the branch can shrink or grow. The Absolute change in resistance is used to decide whether the branch should migrate or stay at its current position. The growth or shrinkage can be identified through the increase or decrease in the resistance during the process.

The updated *health* value of the dendrite branch decides whether to produce the offspring, to die or remain unchanged with its updated *health* value. There are three possibilities of the updated *health* values

- If the updated value is above a certain threshold ($H_{db_{max}}$) value, then it is allowed to produce offspring.
- If the updated value is below a threshold ($H_{db_{max}}$) value, then it is removed from the dendrite.

If an offspring is produced, then a new branch is introduced in CGPDN grid point connected to the same dendrite. It is also the responsibility of the user to specify the values of R_{DB}, $H_{db_{max}}$ and $H_{db_{min}}$.

3.3.2 Life Cycle of Soma

It is also a scalar processing chromosome. This chromosome is used for evaluating the life cycle of a neuron. The output of this chromosome is the updated values of health and weight of the soma. The decision about the soma producing an offspring is dependent upon the value of the health. There are three possible values of the updated health as well.

- If the updated value of health is above a certain threshold ($H_{s_{max}}$) value, then it can produce offspring.
- If the updated value of health is below a certain threshold ($H_{s_{max}}$) value, then it is removed from the network along with its dendrites, dendrite branches and axon branches.

In case it produces a new offspring; then a neuron with random number of dendrites, dendrite branches and axon branches is introduced into a pseudo-random grid location. There are upper (B_{max}) and lower (B_{min}) limits for the random number of dendrites, dendrite branches and axon branches. The model under analysis has an upper limit (B_{max}) of 5 (may be allowed for evolution to decide), and a lower limit (B_{min}) of 2. The soma and branches are provided with an initial value of health, and a pseudo-random value of resistances, statefactors and weights. Users specify the values of the $H_{s_{max}}$, $H_{s_{min}}$, B_{max} and B_{min}. Figure 9 shows the inputs and outputs of the soma life cycle processing.

3.3.3 Axo-Synaptic Branch Life Cycle

It is a scalar processing chromosome whose role is just like the dendrite branch life cycle chromosome. The inputs to this chromosome are the health and resistance of the axon branches, while the output is in form of the corresponding updated values of health and resistance. The decision about the growth, shrinkage or no change is made on the basis of the updated value of the resistance. It can move to a different random neighbouring location if the value of the axon resistance is above certain threshold (R_{AS}).

The updated value of the health decides the future of the branch. There are three possibilities on the basis of the updated value of health:

- If the updated value is above the threshold ($H_{as_{max}}$) value, then the branch will produce an offspring.
- If the updated value is below the threshold ($H_{as_{max}}$) value, then the branch will be removed from the axon.

The production of offspring will give rise to a new branch at the same CGPDN grid point connected to the same axon. Figure 9 shows the inputs and outputs of the axo-synaptic branch life cycle chromosome. Next section explains the information processing mechanism in the entire network.

4 Information Processing in the Network

There are three steps involved in the initialization of information processing in CGPDN. These steps are

(1) Production of a random CGPDN Network with neurons and neurite branches located at pseudo random locations in CGPDN grid.
(2) Production of the initial population of genotypes, with each consisting of seven chromosomes.
(3) The description of the number of inputs and outputs of neural network and their distribution at pseudo random locations in the network.

A pseudo-random network can be formed by specifying:

(1) the initial number of neurons (N_i),
(2) the maximum number of branches per dendrite and axon (N_{bmax}),
(3) the maximum number of dendrites (N_{dmax}),
(4) the maximum neuron state factor (N_{sf}),
(5) the maximum branch state factor (B_{sf}),
(6) the mutation rate (μ),
(7) the dimension of the 2D toroidal space of neuron i.e. the number of rows (N_{row}) and columns (N_{col}),
(8) the neuron and branch life threshold (H_{min}),
(9) offspring threshold (H_{max}), and

(10) the health, weight, and potential reduction factors of neurons and branches ($\sigma_{H_{db}}, \sigma_{H_s}, \sigma_{H_{as}}, \sigma_{W_{db}}, \sigma_{W_s}, \sigma_{W_{as}}, \sigma_{P_{db}}, \sigma_{P_s}$ and $\sigma_{P_{as}}$).

An initial number of neurons having different number of dendrites and axon branches are produced. Every dendrite has a range of branches. The neurons along with their branches are pseudo-randomly positioned in the 2-Dimensional CGPDN grid. Soma and the branches of dendrite and axon are assigned with initial pseudo-random values of health, weight, with branches assigned resistance value as well. The thresholds for all the operations are specified. After the creation of network, initial population of genotypes take place, by stating:

(1) the number of off-springs (λ),
(2) the number of nodes per chromosome,
(3) the set of node functions and the number of connections per node.

The initial number of inputs and outputs of the system and their corresponding locations in the CGPDN grid are also specified at the start. Once the population of genotypes is created, then the rules for information processing in the network are to be specified, as follows:

(1) The input from the environment should be applied to the network first to start of the process.
(2) The network should then run for five cycles (N_{cycles}) (user defined parameter, the value can change) before reading the updated values of output.
(3) A 10% reduction (user defined parameter, the value can change) in the potential of soma and branches($\sigma_{P_{db}}, \sigma_{P_s}$ and $\sigma_{P_{as}}$) should occur after every 5 cycles (user defined parameter, the value can change).
(4) The weights and health of the soma and branches should also reduce by 10% after every five cycles (user defined parameter, the value can change).
(5) The soma threshold potential (η_{th}) should be reduced by twice the reduction (user defined parameter, the value can change) factor of soma potential after every cycle.
(6) The state factor of all the branches and soma should be reduced by one unit (user defined parameter, the value can change) after every cycle which allows them to move towards activity.

The CGPDN network is now ready for operation. To apply the input to the network:

(1) First of all find the location of each input branch.
(2) Select the active dendrite branches at that location.
(3) Bias (duplicate their value in input vectors which is equal to the number of active dendrite branches in this case) the input axo-synaptic branch potential (user defined parameter, the value can change).
(4) Apply the potentials of active dendrite branches and biased input potentials as input to axo-synapse electrical chromosome (CGP program).
(5) The axo-synapse electrical chromosome program will update the values of the dendrite branch potentials (shown in Fig. 10).

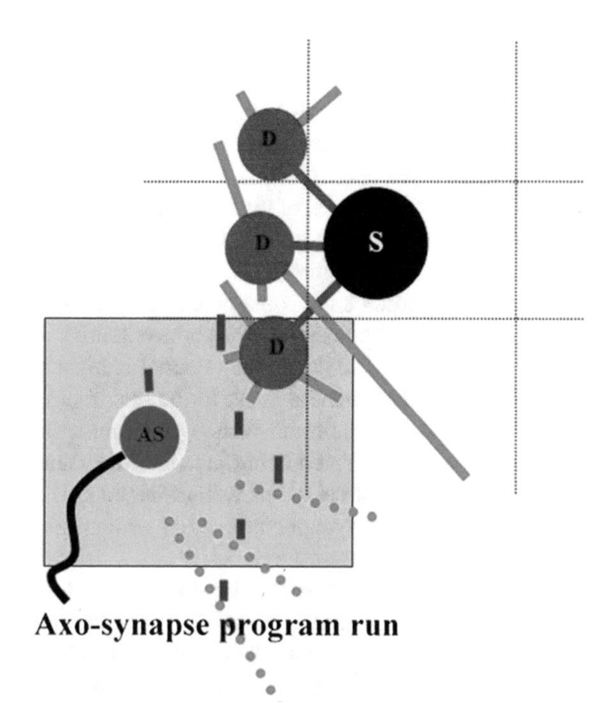

Fig. 10 A diagram of input signal transfer to CGPDN by executing axo-synapse program. The input branch is shown by a black line. The dark circle represents soma electrical processing chromosome (S), red circles represent dendrite electrical processing chromosomes (D), and blue circle axosynapse (AS). The dotted green lines represent dendrite branches of the other neurons in the network. Small red bars at the top of circles and branches show the potentials of branches. Yellow circle highlights the circle whose CGP program is about to run

(6) Repeat the same process in all the input branches.

Once the input is applied to the CGPDN, it must be run for many cycles (5 in this case). In order to run the network for one cycle:

(1) All the active neurons must be selected.
(2) Every neuron is processed one by one in pseudo random sequence.
(3) The processing of every neuron takes place by processing every dendrite connected to the neuron.
(4) At every dendrite, select all the active dendrite branches attached to it. Then apply their potential along with the biased soma potential to the CGP dendrite electrical processing chromosome.
(5) The updated values are produced as outputs.
(6) Process every branch potential on the basis of weight, resistance and health values by using Eq. 1.
(7) After this process, if there is any branch active. Its life cycle is run by applying its resistance, weight and health as input to the CGP dendrite branch life cycle chromosome. This results in updated values. Depending on the updated value of resistance, the decision about migration of the branch from current location is taken. The health decides whether an offspring should be created, the branch should die or remain unchanged.

(8) This process is then repeated for all the dendrites and their corresponding branches.
(9) Once all the dendrites are processed, the average value of potentials of all the dendrites is taken. This is also the average value of all the active dendrite branches attached to them.
(10) The average potential of the active dendrite branches and potential of the soma are applied as inputs to the CGP soma electrical processing chromosome.
(11) This results in an updated value of the soma potential ($\acute{P}$) as output. Equation 2 is used to process the soma potential by using the health (H) and weight (W) of soma. After processing, the soma potential and soma threshold potential are compared. In case the soma potential is higher, the soma fires. This indicates that the soma potential is set to the maximum value. The soma statefactor is also set to its maximum value (Maximum values are specified by the user) which results in soma being inactive for a number of cycles. When the soma fires, its life cycle is run as well as its potential is transferred to other neurons through the axo-synaptic branches by running the CGP axo-synaptic electrical processing chromosome.
(12) In case the soma does not fire, the state factor is adjusted on the basis of the value of the processed potentials. When the soma life cycle is run, the weight and health of the soma are considered as input; and it produces the output in form of their updated values. Once the soma life goes below the threshold (one tenth of maximum in this case), it dies. This indicates the removal of neuron from the network along with its branches.
(13) If the value is above soma offspring threshold, it results in production of another neuron in the network at the same location with a random number of dendrites, branches and other parameters. After the soma fires, the signal has to be transmitted to other neurons which is done by running the axo-synaptic CGP electrical processing chromosome (shown in Fig. 11).

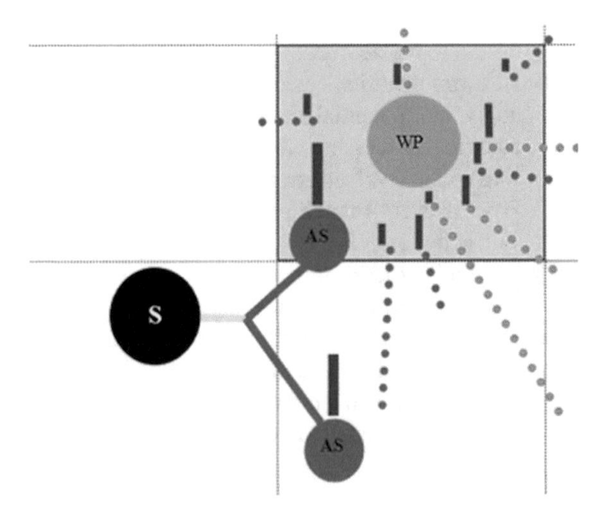

Fig. 11 Axosynaptic potential transfer to the neighbouring dendrite branches, showing Soma(S), two Axo-Synapse(AS) branches, a grid square and a number of dendrite branches attached to other neurons and their corresponding potentials, and a Weight Processing(WP) chromosome. The dotted green (blue) lines represents dendrite (axosynapse) branches of the other neurons in the network

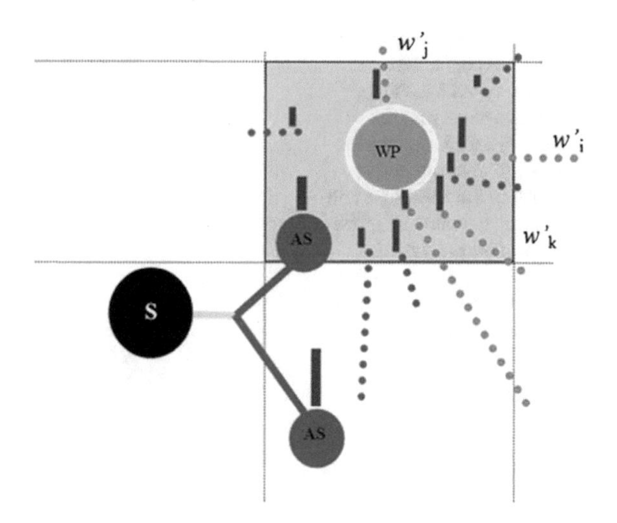

Fig. 12 Weight processing of the neighbouring dendrite branches, showing soma, two axo-synapse branches, a grid square and a number of dendrite branches attached to other neurons and their corresponding weights, and weight processing chromosome highlighted in the grid square. W_i shows the weights of axosynapse and dendrite branches

(14) All the active dendrite branches near every active axon branch are selected and then their potential values along with the biased soma potential (equal to number of active branches) are applied as inputs of the CGP Axo-synaptic electrical processing chromosome.

The chromosome results in updated potentials of all the dendrite branches along with the axo-synaptic potentials. Then Eq. 3 is used for processing the axo-synaptic potential. Once we obtain the processed potential, we run the weight processing CGP chromosome. This takes the weights of the active dendrite branches in the neighbourhood of the axon branch, and its axo-synaptic weight as the input. This also produces the updated values as shown in Fig. 12. The dendrite branch which has the maximum weight after weight processing is assigned the axo-synaptic potential (as shown in Fig. 13). After the axo-synaptic electrical processing, if the potential of the axo-synaptic branch is above certain level; then it is kept active and its life cycle is run. The health and resistance of the axon are the input to the life cycle of the axo-synapse. The variation in the resistance of the branch is compared with a threshold value in order to decide whether the branch should migrate or stay at its current location. If the health of branch is above the offspring threshold, it will results in another branch at the same location. The new branch will have the same health, pseudo randomly selected weight and resistance.

In case the health falls below the threshold value, the branch dies and is removed from the axon. This process is repeated in all the axon branches.

When the network is run for 5 cycles, the output is read from the output branches. The updated potential values of the network processes affect the output branches. When the task is completed, the network fitness is assessed; and the genotype which has the highest fitness is selected. The new offspring chromosomes can be produced through mutation.

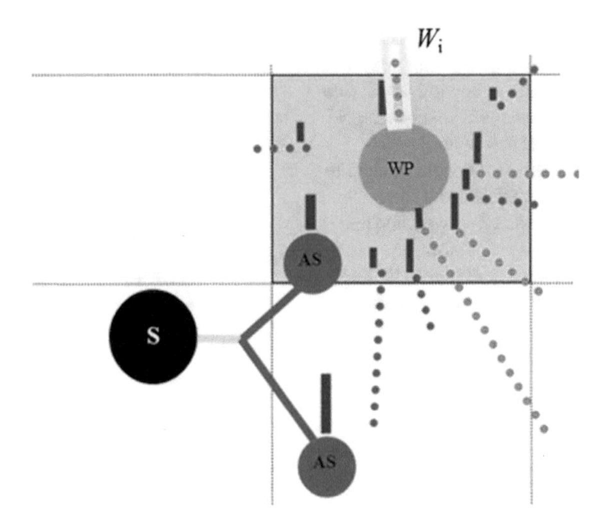

Fig. 13 Transfer of potential to highest weight after weight processing, showing soma, two axo-synapse branches, a grid square and a number of dendrite branches attached to other neurons and their corresponding weights, and a weight processing chromosome, and the highest weighted branch highlighted where the axo-synapse potential is transferred

This chapter described the structure and operation of the CGPDN Model. The inspiration for this model is from the principle of neurosciences. The chapter also presented detailed comparison between ANN models, neural development models and biological neural system. The neurons in CGPDN were arranged in a way that they had a sense of virtual proximity. CGPDN is developed as a result of processing environmental signal by its genetic code. The genetic code is a combinational digital circuit developed by using CGP. The genetic code is evolved by using evolutionary strategies. The evolution is carried out until we get the desired functionality. The input is applied to the CGPDN through the axo-synapse branches by running axo-synapse CGP programs. The output is taken from the output dendrite branches, whose potentials are affected by the CGPDN processes. The inputs and outputs are the interface of the CGPDN with the external environment. Potential is used as a communication parameter between the external environment and CGPDN.

CGPDN is a developmental network which is capable of self-configuration during the task environment. The evolved genotype of the network is executed which results in a complete network of neurons, dendrites, dendrite branches and axon branches, capable of learning. The number of neurons and neurites vary in a task environment. It should also be noted that not all the neurons are active at any one stage of the network.

Next chapter will examine the characteristics and performance of the model in the context of an intelligent agent which tries to solve a learning task known as the Wumpus world.

Chapter 6
Wumpus World

Having gone through all the design procedures, it is now time to test the learning capabilities of the CGPDN in the Wumpus World environment. Wumpus world is a learning problem scenario which is based on agents and artificial environment. All the work discussed over the past few chapters is evaluated in a testing environment to examine and assess the learning ability of the CGPDN. The model discussed under analysis has been tested and results along with the various interesting behavioural characteristics demonstrated are going to be presented in this chapter.

1 Wumpus World Problem

Wumpus world is inspired by "Hunt the Wumpus" game which is an agent-based learning problem (Yob 1975). It is used as a test bed for different learning techniques in the field of Artificial Intelligence. Wumpus World was presented by Michael Genesereth (Russell and Norvig 1995). The Wumpus world is made up of two-dimensional grid and is comprised of many pits, a Wumpus, Gold and an agent. The agent starts from a unique square (home) in the corner of the grid. The agent has to avoid the Wumpus and pits. It also has to find the gold and return to its home. The agent is able to perceive a breeze in squares adjacent to the pits, a stench in the squares adjacent to the Wumpus, and a glitter of the gold square. The agent can also have one or two arrows for shooting the Wumpus. Some of the environments provide a safe route for retrieving the gold. There are also some environments in which the agent has to choose whether it wants to go home empty handed or gamble on either taking gold or die trying. The most common Wumpus world environment is the rectangular grid (Yob 1975 used an environment that was a flattened dodecahedron). Spector and Luke investigated the use of the Genetic Programming for dealing with the Wumpus world problem (Spector 1996; Spector and Luke 1996).

G.M. Khan, *Evolution of Artificial Neural Development*, Studies in Computational Intelligence 725, https://doi.org/10.1007/978-3-319-67466-7_6

1.1 The Proposed Wumpus World

The model introduced in this book has been explored in a Wumpus world environment, however some changes in the environment are introduced to improve the capacity of learning for the agent. The agent is only weakened by the Wumpus and is not killed. The agent doesn't possess arrow intended for shooting the Wumpus. The glitter is on the side squares of the gold. The CGPDN learns everything from the scratch and builds its own memory.

The agent starts with a couple of neurons randomly interconnected. Evolution helps in establishing a stable computational network and finds various ways for processing infrequent environmental signals (Fig. 1).

Evolution helps the agent in navigating around the Wumpus world environment by building the capability of memory development. The goal-driven behaviour in the agent is initiated due to evolution. The Wumpus world for the model of this book is a 10×10 two dimensional grid. It has 10 pits, one Wumpus and the Gold. The location of the pits, Wumpus and Gold is random. The agent has the ability to perceive the glitter, stench and breeze in the adjacent squares. The input to the CGPDN is represented by a number called the potential. The value of the potential is dependent upon what the agent sees. Zero potential indicates that there is nothing on the square. If the potential is 60, it means that there is a pit in the square. 120 indicates the presence of a Wumpus while 200 indicates the presence of the Gold. Magnitude of the input signals to the agent varies depending upon the direction that the agent perceives the signal from. The magnitude of breeze potential can help the agent's CGPDN in perceiving the direction of the breeze. The book chooses the following set of values for representing various directions. North side of the pit is represented by a value of 40, east by 50, south by 70 and west side of the pit is represented by a value of 80. The direction of stench signal is also represented by the magnitude

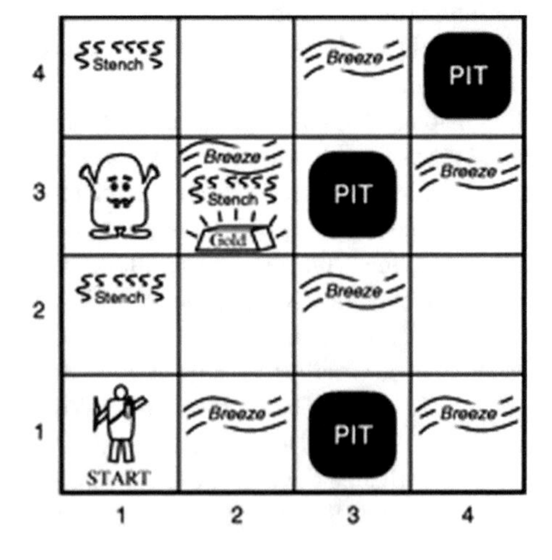

Fig. 1 A two dimensional grid, showing Wumpus world environment, having one wumpus, One agent on the bottom left corner square, three pits and gold (Russell and Norvig 1995)

Table 1 Priority of various signals

Square	Priority order
Wumpus	1
Gold	2
Pit	3
Stench	4
Glitter	5
Breeze	6

of the potentials. North side is represented by 100; East by 110, South by 130 and West by 140. The values for the glitter of gold are 180, 190, 210, and 220 for north, east, south and west respectively. A special input signal helps the agent detect that it is at its home square. Other safe locations do not provide any signal. The agent is able to perceive only one signal in the square even if there are multiple signals available. The priority table for the signals is given below (All these parameters are user defined and purely based on intuition and subject to change).

In Table 1, one(1) means the highest priority while six(6) the lowest. The agent has a quantity called Energy. The initial value of the energy is chosen to be 200 units which reduces by 60% if the Wumpus catches the agent. A pit will reduce the energy by 10 units while gold increases the energy by 50 units. When the agent reaches home, it perishes. With every single move, the agent loses energy by one unit. That is why the agent has to constantly search for energy by solving various tasks otherwise it will die.

The ability of an agent to complete learning tasks can be used to calculate the fitness of agent. The fitness keeps accumulating over the period where the energy level of the agent is more than zero or before it returns home. The fitness value can be calculated when energy $E >= 0$. With every move of the agent, its fitness increases by 1 unit to encourage the agent to have a brain; which remains active and does not die. When the agent returns home with gold, its fitness increases by 2000; while returning without gold will increase the agent's fitness only by 200. In case the agent obtains gold its fitness increases by 1000.

The Wumpus world game starts when, the potential used to represent the status of square that is Home is fed as the input to the agent. The computational system of the agent receives the input through the axo-synapse branches that are distributed across the CGDPN at five distinct locations; thus updating the potentials of neighbouring dendrite branches in the CGPDN grid linked to different neurons.

Then the network operates for five cycles which updates the output potential of dendrite branches. The potential of dendrite output branches is also the output of the CGPDN. Once this step completes the updated potentials of all the output dendrite branches are recorded and then averaged. This average potential is utilized for assisting in the decision regarding the direction of movement of the agent. For more than one direction, the potential is divided into as many ranges as the possible directions of movement.

The agent dies when either it runs out of energy, or the neurons die, or all the dendrite or axon branches die or the agent gets home safely. There are five (1+4 evolutionary strategy) randomly generated genotypes produced. The fitness is obtained by assessing the corresponding agent behaviour. The best genotype is chosen as parent for a new population.

1.2 CGPDN Setup

The book uses a 3×4 CGPDN grid (it is different from the Wumpus world grid). The smaller grid helps the branches communicate with each other because of the smaller number of neurites and neurons. The inputs and outputs of the network are positioned at five different random CGPDN grid squares. The initial number of neurons is set to 5, the dendrites, the number of dendrite and axon branches are all pseudo-randomly generated with upper limit being 5. The initial structure of the network is selected pseudo-randomly which adjusts itself in the task environment. The computation time is reduced by letting the maximum branch statefactor to be 7. The statefactor of soma is kept at 3 as it should be inoperative after firing for a minimum of two cycles. The rate of mutation is set at 5%. The quantity of nodes per chromosome assigned has a maximum value of 100. 8-bits are used for representing the potential, resistance, weight and health for this model. Its maximum value can be 255. The length of chromosome is 400 integers ($100 \times$ (3-inputs per node + node function)). All these parameters are intuitively chosen and are subject to change.

1.3 Results and Analysis

The agent's performance is assessed on the basis of its ability to solve three types of tasks. These tasks are

- Finding its way back home.
- Finding the gold.
- Bringing the gold back to home.

The agent must perform all these tasks in one wumpus world independently within a time frame of one life. The greatest fitness is attained by performing the last task. The population performance of every agent is evaluated in the Wumpus world where the position of Gold, pits and Wumpus are fixed. The agent struggles to live by evading pits and the Wumpus and return home before it runs out of energy on which its life span depends. An agent must maintain the neural network for solving its task. The model has evaluated the agent's performance in individual evolutionary processes with identical Wumpus world and disparate primary populations.

Table 2 presents the performance of the agent in twenty independent evolutionary runs. The table clearly shows that the agent solves the first two assignments swiftly

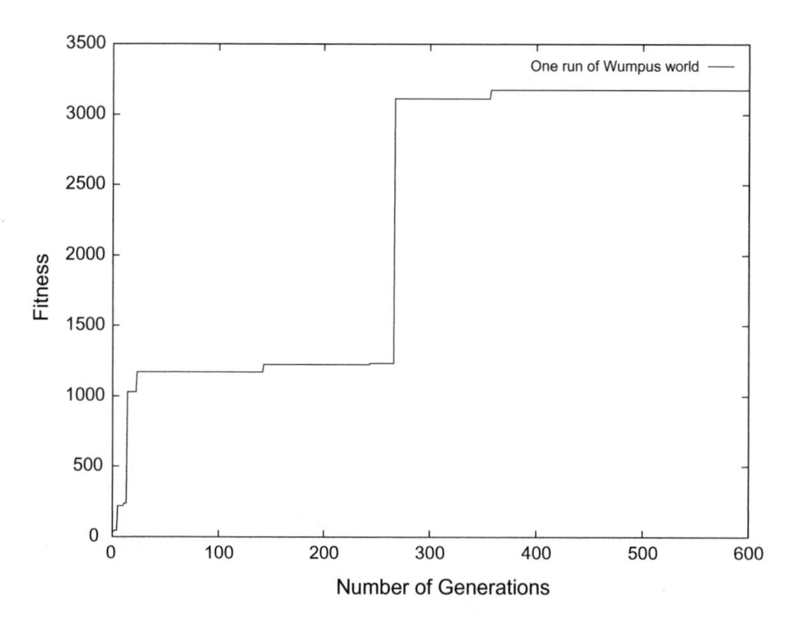

Fig. 2 Fitness diagram depicts the performance of an unrivaled agent over a span of several generations during its evolution

Table 2 The average, maximum, and minimum number of generations taken to perform various tasks in a fixed Wumpus World in 20 independent evolutionary runs

S. No.	Task	Average No. of generations	Minimum No. of generations	Maximum No. of generations
1	Home empty handed	4	1	15
2	Found the gold	13	2	95
3	Home with gold	300	12	1108

compared to the last one as it is much difficult in comparison. Figure 2 depicts the fitness of an agent during its training process to solve the Wumpus world problem. The fitness is initially low, but after a few generations (12); the fitness curve shows a drastic increase when the agent manages to find the gold. The fitness of the agent keeps increasing as its life span increases and is successful in finding gold. After 280th generation, the agent found the route to home with the gold.

The network is evaluated further with similar primary population and different Wumpus worlds where the location of pits, Wumpus and gold were changed. Diverse agents are evolved in every run which are able to solve these tasks. Table 3 highlights the effectiveness of the agent in ten individual evolutionary runs. The table makes it clear that, similar number of generations are recorded as the first case for solving the problem. This indicates that the initial Wumpus world generated arbitrarily had an average level of toughness.

Table 3 The average, maximum, and minimum number of generations taken to perform various tasks on different Wumpus Worlds starting with the same initial population in ten independent evolutionary runs

S. No.	Task	Average No. of generations	Minimum No. of generations	Maximum No. of generations
1	Home empty handed	5	2	11
2	Found the gold	20	2	79
3	Home with gold	530	12	1769

The fitness function is arranged in such a way that initially it forces the agent to sustain its network which results in an increase in its life span. Initially the agent does not achieve any goal, but fitness improves with time which agent spends in the environment. During this time, the agent has to maintain the network; that's why evolution attempts to shape a steady and robust computational framework in the earlier part of evolution. After achieving this, evolution initiates the production of agents that learn how to come back home, how to find the gold and finally, how to bring the gold to home.

Initially, the agent does not know anything about the Gold, Home, Wumpus, Pets and the signals which indicate the presence of these subjects. When the system is evolved, the genetic code in the agent allows building a computational structure which holds memory of the meaning of these signals during the agent's life cycle. The source of building this knowledge is the initial genetic code when it is run on the initial randomly wired network. After examining the various runs of the experiment, one can conclude that in all the cases; agent learns to avoid the Wumpus, pits and strives to get the gold. In CGPDN, those programs are evolved which can build and change continuously a computational network within an external environment.

Conventional artificial neural networks operate by updating weights in order to converge towards the solution of a particular assignment. These networks require retraining for a minute variation in the essence of the problem (Cunningham et al. 2000).

One of the major points of interest is, whether these evolved programs can build a network that can lead to successful agent behaviour in different Wumpus worlds. To explore this, we chose the program of the more evolved agent showing promising results in a particular Wumpus world and test its potential in randomly generated Wumpus worlds. Various tests indicate that by changing the locations of the pits and Wumpus, the agent is able to always bring the gold to home. However, if the position of gold is changed; then some of the agents can get the gold (30% of the time) but they cannot find the route to home. About 50% of the time the agents returned home without gold. 20% of the time the agent was neither able to find the gold nor able to return home. These facts indicate that the model has to be improved as there is still some noise in the network. We need to achieve a general problem solving behaviour.

One of the interesting findings of these tests is that the agent always looks for the previous position of the gold even if the environment is changed. This is a positive indication because it shows that the evolved genetic codes can build a computational network which is capable of retaining information.

1.4 Development of Network Over Agent's Lifetime

The Structure of CGPDN changes while it is solving the Wumpus world. Figure 3 demonstrates the changes in the energy of one of the agent while searching for gold and bringing it back home. The energy level decreases after the agent falls into the pit. The energy decreases by 10 units for every fall. The figure shows the case where agent falls into the pit three times. However, the Wumpus does not catch it. Once the agent finds the gold its energy increases by 50 units. After the agent reaches home, it is terminated by making its energy equal to zero.

Figure 4 depicts the variation in the number of neurons, dendrite branches and axon branches during an agent's life while Fig. 5 demonstrates the variation in the number of active neurons, dendrite branches and axon branches at various stages during the

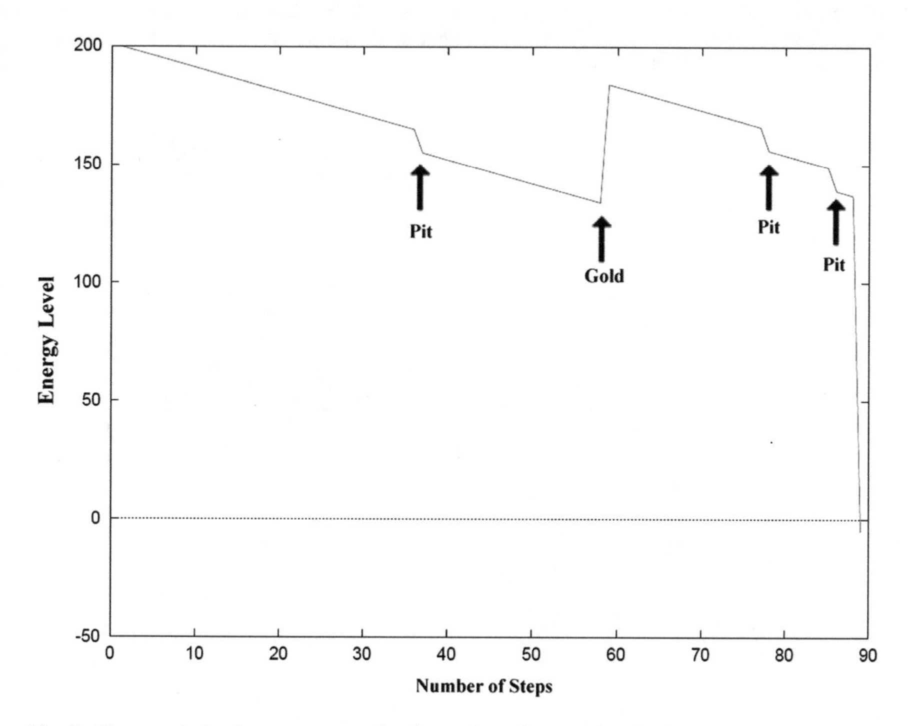

Fig. 3 The correlation between energy level, number of steps taken by the agent and variation in its energy level on encountering gold or a pit is shown

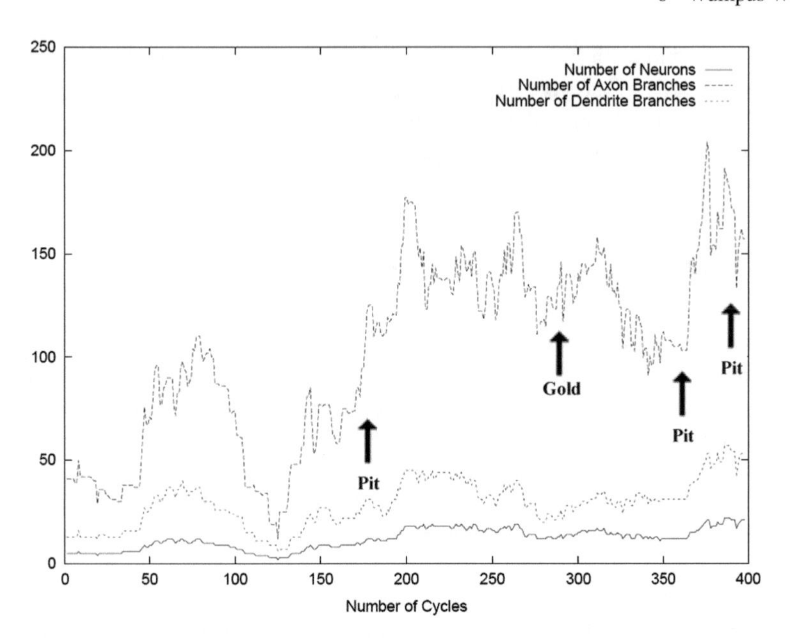

Fig. 4 Variation in numbers of neurons, and neurites (axon branches and dendrite branches) at various stages of an agent's life while solving wumpus world problem shown against the number of completed cycles. The numbers of these components fluctuate at the start, and then appear to stabilize later at around 200 cycles

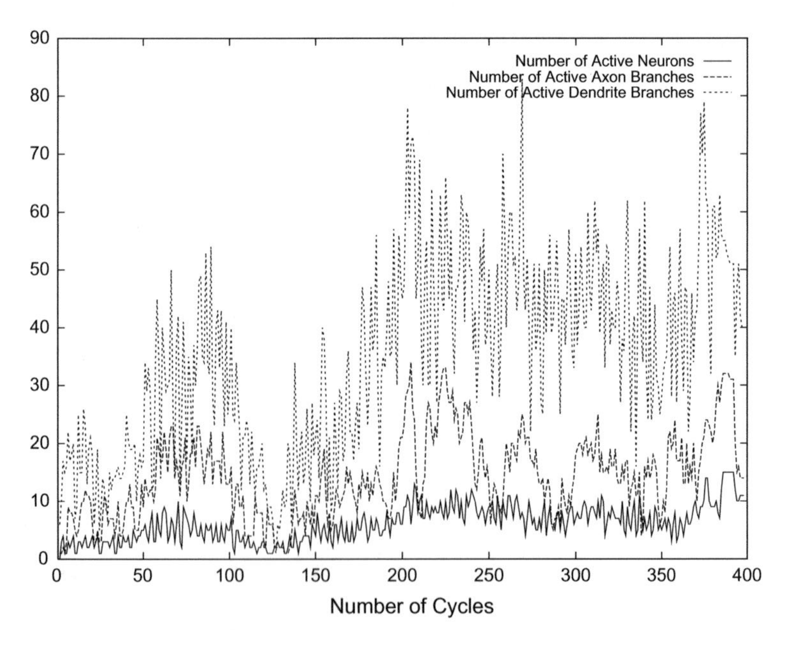

Fig. 5 Variation in numbers of active (with state factor zero) neurons, axon branches and dendrite branches at different stages of an agent's life while solving wumpus world problem

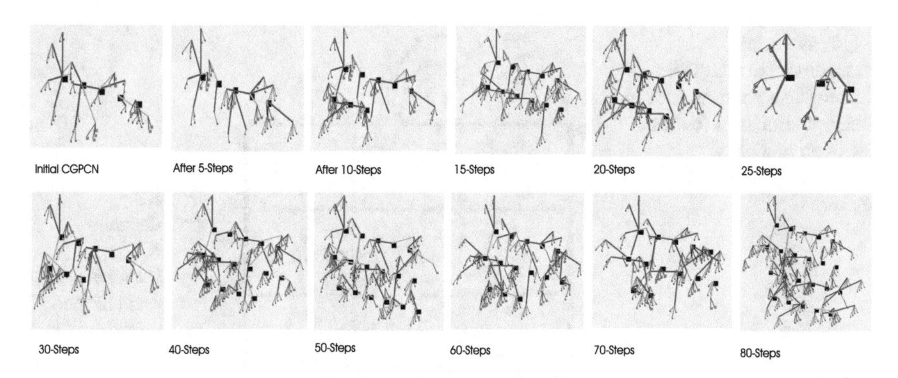

Fig. 6 Structural changes in agent CGPDN network at different stages of wumpus world shown against number of completed steps. The network starts with 5 neurons and end up with 21 neurons after completing 80 steps. Black squares show somas, red thick lines the dendrites, yellowish green lines the axons, green lines the dendrite branches, and blue lines showing axon branches

life time of agent. Figure 6 shows the changes in the neural structure and growth of CGPDN at different stages. With the development of the network, old dendrite branches disappear; while new dendrite branches arise. After ten steps, the network gains three new neurons. Figure 6 also shows that network changes randomly, but it adopts a stable topology later. An interesting fact is that the network starts with 2–4 dendrites and after all the variations that take place, it ends with the similar numbers.

Figures 4, 5, and 6 present very fascinating network dynamics. The network dynamics are random initially which then get a robust framework as the agent performs its task. The time taken by the agent to achieve its second goal is way lesser than the time it takes to achieve the first goal. The reason behind this is the development of a sustainable network which allows it to quickly achieve the goal.

This also suggests that an agent can make a map of the environment during the development stage, which is used for finding the return path. The analysis also indicates that "the return to home" is usually more direct compared to the path leading to gold. If we visualize that the movement of the agent is determined by throwing a dice, it might take an infinite time to get the gold and bring it home. Since every empty square gives the same signal, the agent might oscillate between any two squares all the time. The pits also give similar signals, so it is hard to figure out the locations of the pits. All these issues make Wumpus world a tough problem to solve.

1.5 Testing the Network Without Life Cycle Programs

We investigate the behaviour of the evolved networks on a Wumpus world which has different initial populations but a fixed morphology in the CGPDN. Fixed morphology means that there is no life cycle program run inside the neurons, dendrite branches and axon branches. That is why the network structure does not vary, and

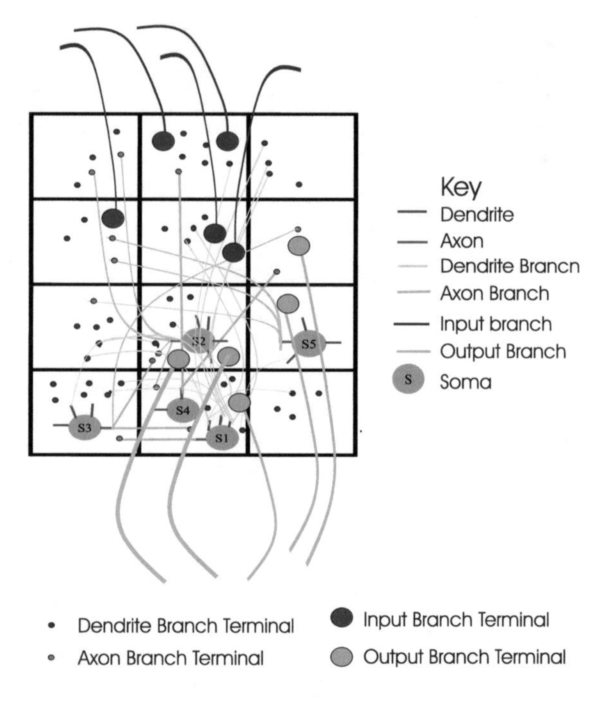

Fig. 7 Neuron's arrangement in CGPDN grid showing five neurons with different structures (dendrite and axon branches) representing the exact locations

only electrical and weight processing chromosomes are run and evolved. The model under analysis is also provided with a random CGPDN structure which initially has five neurons.

First (S1) and second (S2) Neurons are provided with four dendrites; neuron three (S3) and five (S5) with three while neuron four S4 with one dendrite as depicted in the Fig. 7. The position 5, 2, 5, 2 and 4 in the CGPDN grid are provided with a distribution of axo-synaptic branches. At location 8, 6, 11, 8 and 9 in the grid, the output dendrite branches are assorted. Neuron S1 has one axon branch, neuron S2 has four, neuron S3 has 3, neuron S4 has two and the fifth (S5) neuron has three axon branches.

The arranged neurons are shown in Fig. 7 in a way that they are fully connected. All the neurons are connected in the path starting from the input to the output. This fixed network takes at an average three times longer for solving the Wumpus world problem in comparison to the one in which neurons and neurites has a life cycle and develop in real time. The model discussed in this book is run for four independent evolutionary runs. Table 4 shows that the number of generations which the fixed CGPDN network took to complete various tasks.

The table demonstrates that the life cycle of the neural components is an important factor in the learning ability of the network, since it took longer to achieve some of the tasks, and could not achieve the main task of bringing the Gold home in some of the cases. Chalup proposes an incremental learning system which assists the development of the network during the learning phase. It performs much better in

Table 4 Number of generations took agents with fixed CGPDN to perform various tasks in Wumpus World starting with different initial population in four independent evolutionary runs

Run. No	Found the gold	Home with gold
1	4	290
2	1	1420
3	2	3302
4	0	More than 5000

comparison to artificially imposed fixed network structure (Chalup 2001), supporting the above argument.

1.6 Learning and Memory Development in CGPDN

The model in this book is further evaluated through various experiments performed on the Wumpus world. It is also exploited to check the validity of the argument stating that "the genotype of agent obtained through evolution holds a memory of its ancestors' history of the Wumpus world environment in its genetic structure."

Experiments are conducted to evaluate the learning behaviour of ten highly evolved agents in various scenarios. Instead of killing the agent after it reaches home with gold, it is allowed to move in the environment and live longer. 50% of the time the agent went towards the gold, while at other times the agent moved randomly in the environment and then came back home. The agents then go out again and visit different areas. They return back home and then end up oscillating near the home (30% of the time). 20% of the time, the agent leaves home and gets stuck somewhere in the corner of the Wumpus world where it oscillates and eventually dies. Mostly the agent strives to get the gold again while finding a shorter path to reach the gold. This indicates the presence of the map of the environment in the genetic code of CGPDN. It also indicates that the agent has a goal driven nature, which is why it goes for gold again and again. Some agents ended up showing oscillating behaviour which may be caused by the agent being in an unusual situation of being in the corner. However this is a rare behaviour.

We have examined a specific case in which an evolved agent retrieves the gold three times. However it ends up oscillating. The evolved agent reaches the gold by following the path shown in Fig. 8a and then gets back to home by utilizing the route shown in Fig. 8b.

Figure 10 demonstrates all the variations in the energy level of the same agent. Figure 9 shows the steps of agent across grid squares in terms of grid numbers.

It is interesting to know that the agent that moves in a straight line is highly non-random. The probability of a random move at the edge is 33%. So the probability of an eight step straight path while returning to home is 0.0152%. The probability

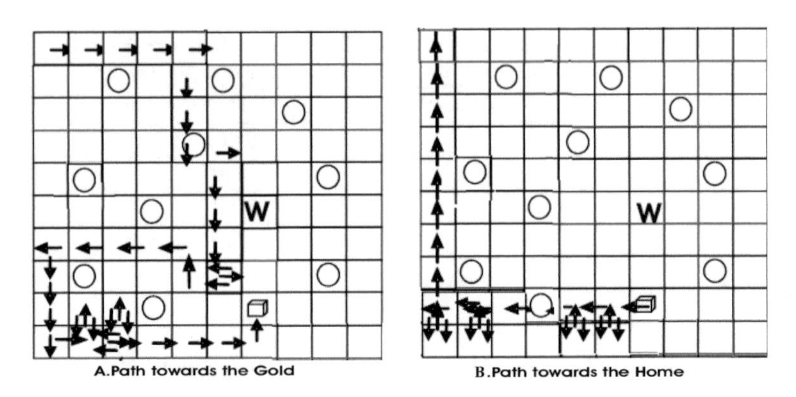

Fig. 8 Path followed by the agent, starting from upper left corner (home square), **a** from home towards the gold, **b** from gold back to home

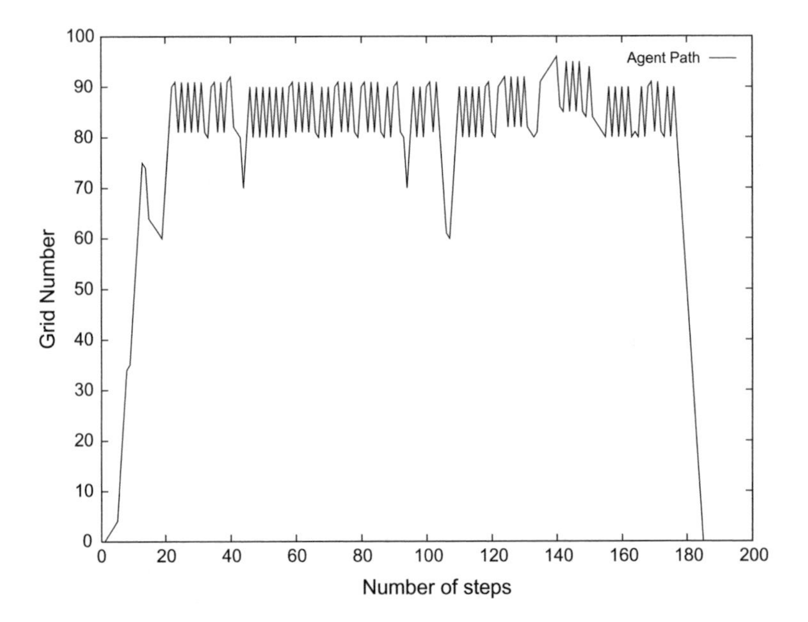

Fig. 9 Path followed by the agent in graphical form, from home to gold and back to home, with grid points enumerated along rows from top left to bottom right (0–99). The graph highlights the oscillatory behaviour of the agent

shows that it is very unlikely to be random. The work then examines the movement of the same agent. Figure 8 shows the various paths which the agent follows in four subsequent journeys. Figure 12 shows the changes in the energy level of the agent. There is a similarity between Figs. 11a and 8a which shows the first path followed by the agent towards the gold. During the first journey, the agent takes 135 steps to find the gold. Figure 11b shows the path followed by the agent in its second journey.

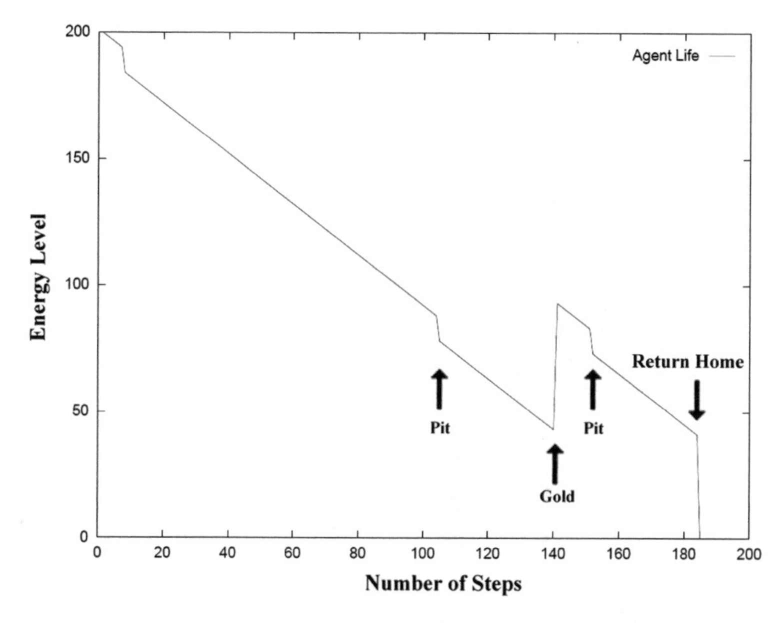

Fig. 10 Variation in agent energy during its trip from home to gold and back to home. Their is a continual decrement in energy of 1 per step, and sudden drops when caught by pit (steps 7, 104, 151), and a sudden increase when it gets the gold (step 140)

It can be seen that in this journey the agent almost takes a straight path towards the gold with a few oscillations on the way. The agent also encounters a pit on the way which is shown by the dip in energy level in Fig. 12. The agent takes 38 steps for the second journey. Figure 11c demonstrates the paths taken by the agent in its third journey. The path that the agent chooses is almost same as the second journey, however this time it is with more oscillations. It also faces two pits. It reaches the gold in a longer time compared to the second journey. It took 44 steps for the third journey. The fourth journey also has a similar path like the third journey, however this time the oscillations increase and the agent gets stuck in a corner until its energy level becomes zero, and it dies.

2 Competitive Learning Scenario

Co-evolutionary computation is largely used in a competitive environment. In competitive evolution, the fitness of individual is based on how it competes against an opponent. Hence fitness is the measure of the relative strength of solution. Such competing solutions results in an "arms race" of increasingly better solutions (Dawkins and Krebs 1979; Rosin and Belew 1997; Van Valin 1973). There is a feedback mechanism present between the individuals on the basis of their selection, which produces

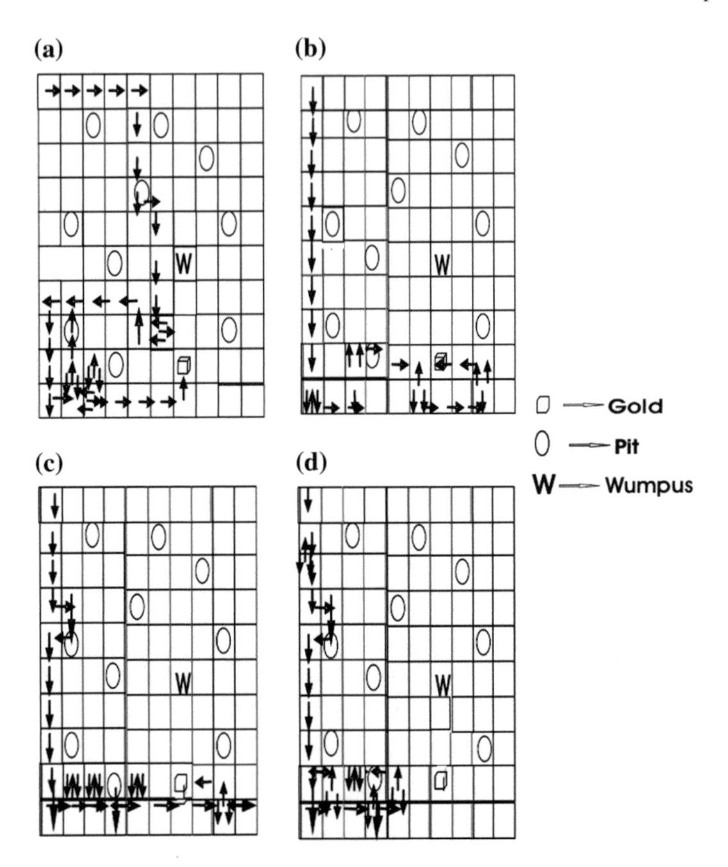

Fig. 11 Trips of the agent in four scenarios in search of gold. **a** Primary trip to find gold beginning with an initial random CGPDN. **b** Second trip taken by agent to take the gold home, note the relatively shorter route taken. **c** The agent attempts to find the shortest path to acquire gold but ends up getting caught twice. **d** A trip similar to that of case c is followed, but the agent begins to oscillate and ceases by losing its energy

a strong force towards complexity (Paredis 1995). Traditionally competitive evolution is used for evolving interactive behaviours which is a hard task to evolve using an absolute fitness function. For encouraging interesting and sophisticated strategies in competitive co-evolution, every player's network should compete against a high quality opponent.

There is a feedback mechanism present between the individuals on the basis of their selection, which produces a strong force towards complexity (Stanley and Miikkulainen 2004). Traditionally competitive evolution is used for evolving interactive behaviours which is a hard task to evolve using an absolute fitness function. For encouraging interesting and sophisticated strategies in competitive co-evolution, every player's network should compete against a high quality opponent.

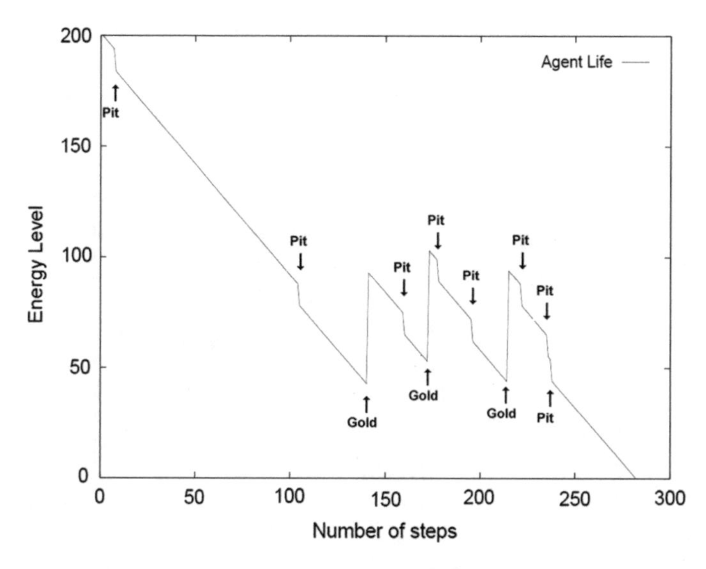

Fig. 12 Variation in agent energy level during the four journeys illustrated in Fig. 11. The energy shows a continual decrement of 1 per step, sudden increases when the gold is retrieved (steps 40, 172, 214), and sudden drops when caught by a pit (steps 24, 104, 159, 178, 195, 221, 235, 237)

The work in this book also evaluates the performance of the CGPDN in co-evolutionary scenario, in which both agent and the Wumpus are provided with a CGPDN. The agent and Wumpus live in a 2-D 10×10 grid which has 10 pits as shown in Fig. 13. Here the Wumpus is also provided with a home square which is located at the bottom right corner as shown in the Fig. 13. The agents' task is slightly modified. Once the agent gets the gold, it automatically gets to home. It is then sent to get the gold again. The task of the agent is to get gold as many times as possible during its life time while avoiding the pits and Wumpus. Wumpus has to catch the agent as many times as possible. The task of the Wumpus is difficult compared to the agent since the target of the agent is static while the target of Wumpus is mobile.

Both the Wumpus and agent are placed at their home squares at the start of the experiment. Learning is a difficult task in the co-evolutionary scenario compared to the single agent world. The reason behind this is that every time the Wumpus catches the agent; its task of catching the agent again becomes harder, because the agent learns and tries a different path to the gold in order to avoid the Wumpus. Pits affect the energy of the agent. There is a decrease in the energy of the agent when it passes through the pit. The home of the Wumpus is diagonally opposite to the home of the agent. The agent and Wumpus can sense the breeze in squares adjacent to the pits while they sense the smell of each other when they are in squares adjacent to each other. They can sense the glitter while passing near the gold. They receive different signals while passing through squares near these locations. The agent and Wumpus have to learn not only how to deal with the breeze and smell, they also have to sense the direction from where they are coming and make a decision to move accordingly.

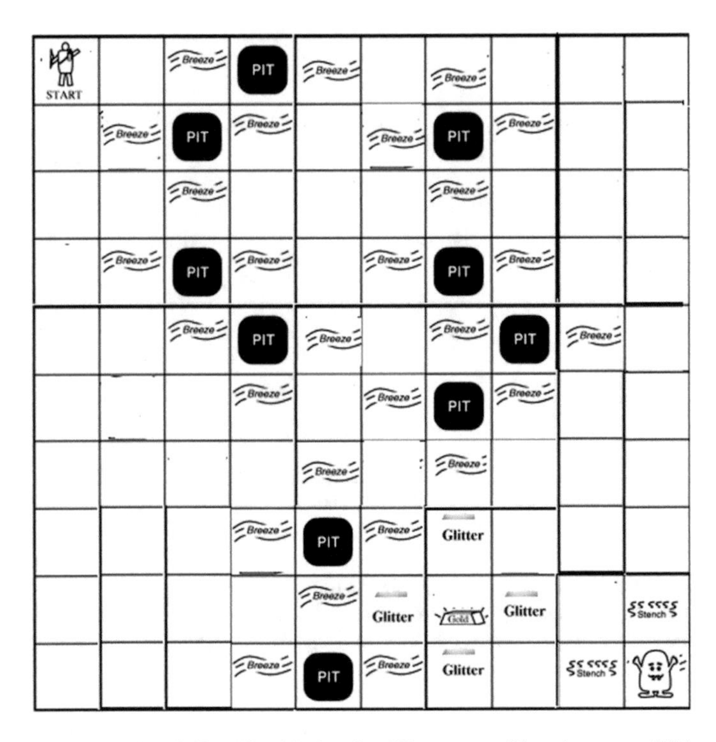

Fig. 13 A two dimensional 10 × 10 grid, showing Wumpus world environment, With a wumpus at bottom right corner, an agent on the top left corner, ten pits at various places and gold on square 86

All the safe locations other than the home provide zero signals. The pits and the gold directly affect the agent, which is why the Wumpus has to learn how to differentiate between all these noise signals. It also has to identify the presence of the agent for catching it. Both agent and Wumpus can perceive only one signal on a grid square and stay alive as long as their energy is above "zero".

Initially both the agent and Wumpus have the energy level of 100 units. They also have only one life cycle. They can move only one square at a time. They cannot stay in the same square, since they must move in any of the possible directions. In case the Wumpus catches the agent, its energy reduces by 60% while the energy of Wumpus increases by 60%. With every single move, the energy of both the Wumpus and the agent decreases by one level. Their fitness keeps accumulating over their lifetime. With every move, their fitness increases by one. When the agent is successful in finding gold, its fitness increases by 1000. When the Wumpus catches the agent, the Wumpus's fitness increases by 1000.

The behaviour of five agent population members against the best performing Wumpus genotype from the previous generation is explored. The initial random network remains unchanged for both the agent and the Wumpus. The genotype in each generation is responsible for generating different networks and their associated functionalities. The best agent and Wumpus genotypes are selected as the respective parents for the new population.

It is demonstrated that the agents learn during their life time while their energy level is above zero. This showcases that evolution enhances the ability to learn. The initial CGPDN setup is the same for both the agent and Wumpus. Thus the genetic code in the mature network builds the genetic memory which is obtained by the agent during its life time which can be transferred to next generation through the genetic code. Only genotype changes in each generation. The starting neural structure of both the agents is same.

2.1 Results and Analysis

Figure 14 demonstrates the change in the fitness of agent and Wumpus in one particular evolutionary run over 1250 generations. It is clear that there can be increases and decreases in the fitness at various stages of every agent and Wumpus where either of them can hold the edge over the other.

The dynamic behaviours of the agents and Wumpus vary. These are the behaviours that the Wumpus and agent adopt due to the interactions between the neurons. They

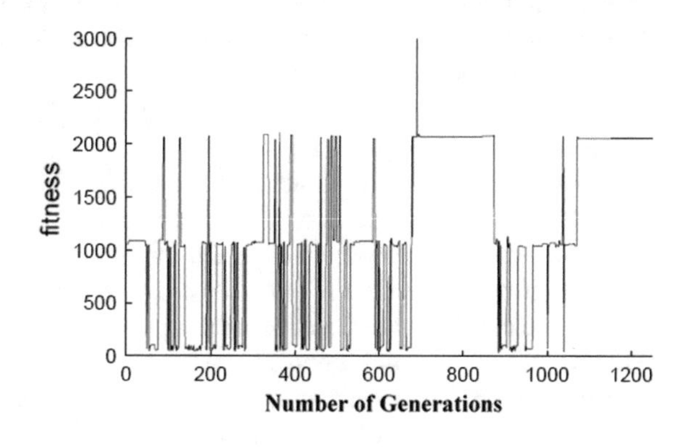

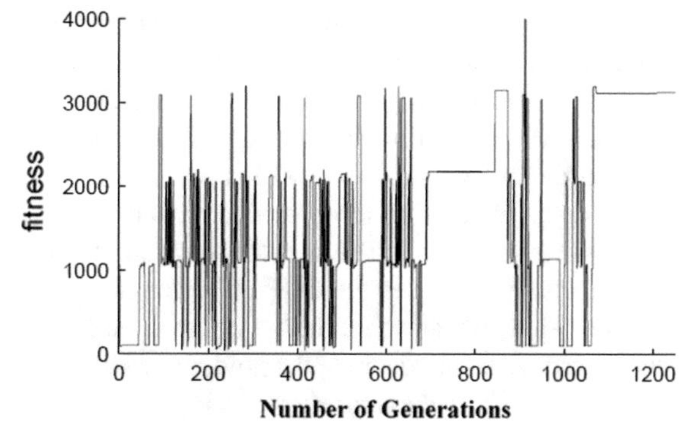

Fig. 14 Variation in fitness of the best agent (top) and wumpus (lower) during evolution. Oscillatory response in fitness variation of both at the beginning, with improvement in fitness at a higher generation

arise due to the internal genetic code. Initially the agent and Wumpus are not aware of the gold, pits, their interactions with each other and the signals which indicate the presence of the nearby objects. As they evolve, they develop their own memory of their life's experiences which is encoded genetically. Initially the agent has a randomly assigned neural structure which develops during the lifetime of the agents, making it reasonably skilful and capable to achieve the goal.

The fitness graph in Fig. 14 shows that initially the fitness of the agent and Wumpus varies a lot. However, with the increase in the number of generations; both the agent and Wumpus become more skilful and the frequency of the changes in the fitness reduces. This indicates that the two CGPDN behave in a manner in which one of the CGPDN benefits at the cost of the other. There are certain points in the graph where the fitness of both Wumpus and the agent goes down; this indicates that both have failed to achieve their tasks. Usually they are able to find a way to achieve their goals in the next generation. It is observed that around generation 680, both agent and Wumpus become reasonably skilful.

The agent obtains the gold twice while on one occasion it obtains the gold three times. The Wumpus catches the agent twice and then three times. Then the agent and Wumpus get a fluctuating fortune. Then around at generation 1100, both the agent and Wumpus are more successful in achieving their goals, however Wumpus is more successful.

While observing the behaviour of the agent and the Wumpus, we observe that, when the initial energy of the Wumpus and the agent is increased to 300, it allowed the agent to find the gold more frequently as shown in Fig. 15. The Wumpus caught the agent just once. Figure 15 also shows the movement of the agent and Wumpus over the 6 journeys. The Wumpus dies during the fourth outing in panel D that is why it is not shown in panel E and F.

The squares in the Fig. 15 are numbered from 0 to 99 along rows from top left to bottom right. The gold is on square 86. In Fig. 15a the Wumpus and the agent begin with the same randomly assigned initial networks. These networks turn into mature networks when the seven CGP programs are run. The agent almost takes a direct route to the gold while encountering two pits on its way towards its gold. The Wumpus spends a lot of time in square 98, 88 and then moves towards the gold just after the agent re-spawns from its home square after obtaining the gold. Figure 15b shows the next journey of the agent, which takes a different, but shortest path to the gold while avoiding all the pits. In the meantime, the Wumpus lurks near the gold, spending all its time on square 96 and 86. This is an efficient strategy as the agent eventually comes close to the grid and this gives the Wumpus a chance to catch the agent. However this time the Wumpus is unlucky and it moves away from the gold at the same time when the agent obtains it.

The agent always has to move since it is not allowed to be stationary. Figure 15C shows the third journey. The agent follows an identical path to the gold as it initially did. This is surprising as the CGPDN network develops with experiences during its lifetime. This indicates that the agent has encoded the map of the environment, but when it arrives at the square of the gold, the Wumpus attacks it and it is relocated to its home square. The behaviour shown in Fig. 15D is interesting. The agent now follows

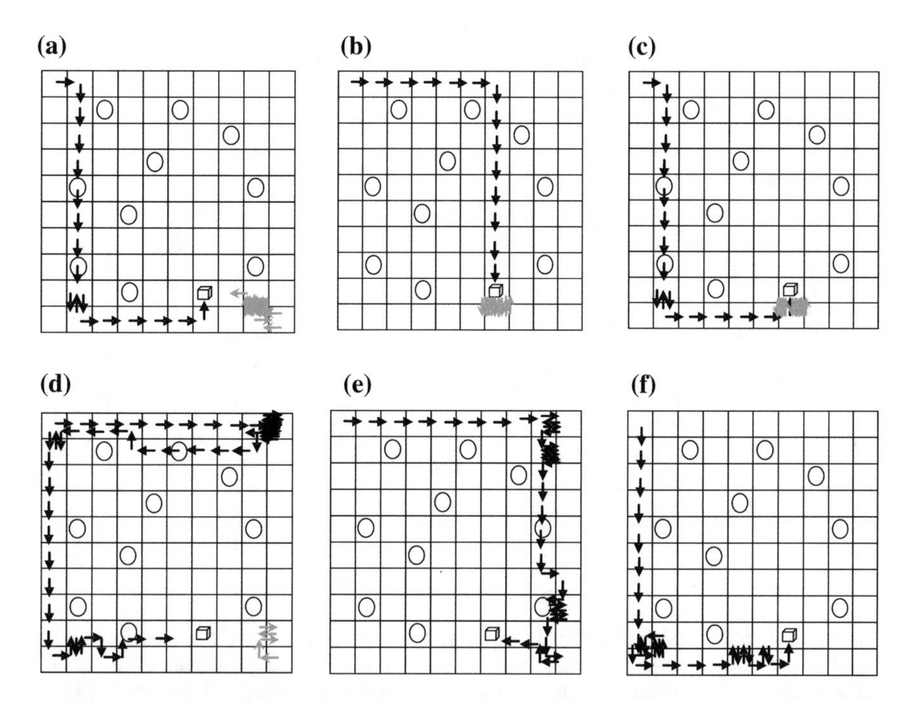

Fig. 15 A–F Various paths taken by agent and wumpus towards their goals in coevolutionary task. The agent starts from home (the upper left corner) towards the gold, the wumpus starts from the lower right corner (home) towards the agent. The agent trip is shown by black arrows, while that of the wumpus by grey arrows, pits by circles, and gold by a box. Wumpus dies during trip **D**, thus not available in **E** and **F**

a different and meandering path to the gold. It spends some of its time alternating between the square 8 and 9 and then the agent turns back and arrives at it home. After which, it goes down the left hand side in the direction of the gold.

The Wumpus, on the other hand behaves in an odd manner. After it attacks the agent, it is relocated to its home. It moves around briefly in the bottom right four squares before its CGP developmental network dies. This demonstrates a very fascinating but strange phenomenon, which we observed with the other evolved agents and wumpuses. Usually, their CGPDN dies when their energy level is a small number and becomes more active when their energy level increases. This is puzzling because the energy level is not supplied as an input to the CGPDN. The beneficial encounters usually result in greater number of neurons and branching, while deleterious encounters usually result in removal of neurons and branches from the network. Figures 15E and F show the subsequent behaviour of the agent in which it obtains the gold again (Panel E) and dies before it reaches the gold (Panel F). These results indicate that the agent can produce a memory map of the environment early in its evolutionary history. Once the Wumpus attacks the agent, the CGPDN of the agent is strongly affected due to which it follows a different path. In case the agent does not find gold, it returns

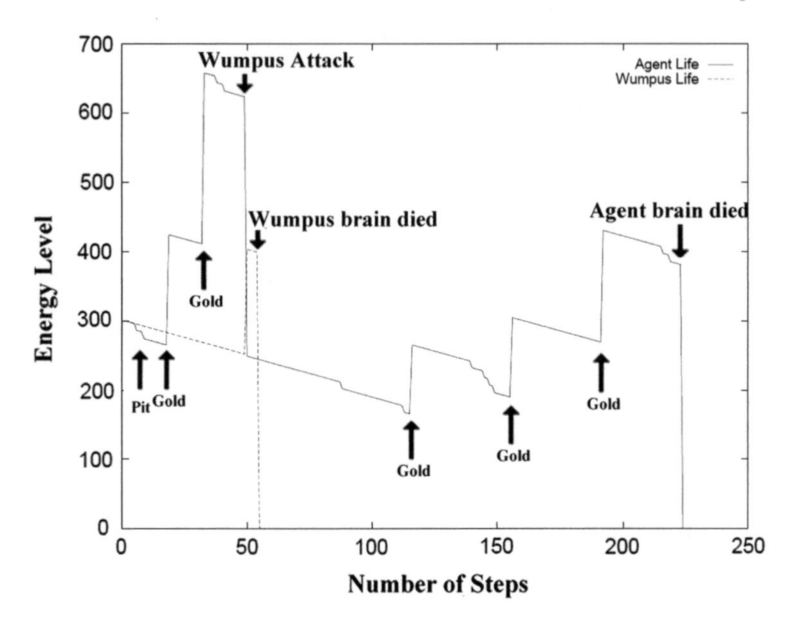

Fig. 16 Fluctuation in the energy level of the agent and wumpus during the task environment of Fig. 15. Continuous drop of 1 per step, and rise of 60% on acquiring gold, sudden drop of 60% when caught by the wumpus, and a reduction by 10 units when encounter a pit. The agent seems to be successful in acquiring the gold 5 times and is caught once by the wumpus

to its home square and then uses the same path which led to the attack of Wumpus. Hence it can eventually find some gold. Further test results demonstrate that the agent looks for the shortest path to gold. Even after the events described in Fig. 15E, experiments revealed that whenever the agent is relocated to its starting position; it follows a short path to get to the gold. But unfortunately on its way to find gold, the agent dies. The behaviour of this agent is evaluated in different situations. When all the pits are removed the agent started to move around the environment apparently at random without finding the gold. So the agent has environmental cues to navigate. The Wumpus is also moved to square 56 which is right in the path of the first agent in Fig. 15B. The behaviour of the agent is similar to its previous behaviour with the exception of getting caught by the Wumpus. It also does not avoid that square when it encounters the smell signal on square 46. This is the indication that this agent's network building program has not yet given it a general response of avoiding the wumpus. Moreover, this encounter also affects the network and causes it to follow a different path in order to avoid the wumpus. Here the agent and wumpus from the generation 220 were explored, since they are not highly evolved, they still have to learn how to behave in the wumpus world environment. The agent does not properly respond to the presence of the wumpus and the degree to which it affects its energy level. If the agent is evolved for a longer period of time, both the wumpus and agent get better and will result in a stable behaviour where the agent gets the gold many times, while at times the wumpus catches the agent as shown in Fig. 14.

In the earlier generations, the agents wander around in the wumpus world environment attaining low fitness values, and not selected during the course of evolution. Initially the cause of increase fitness is avoiding pits and wumpus and live longer, causing it to attain an oscillatory behaviour while lurking between two squares. The rise in fitness occurs when the agent achieves any of their goals during the course of evolution causing them to be selected for the next generation. This is where the agent gets the sense of goal, and the sense of goal makes one of them the agent and other the wumpus. Figures 15 and 16 show the results of well evolved agents which have a sense of goal.

Considerable variation takes place in the structural development and activity of the CGPDN during the life time of the agent and Wumpus. The results of experiments performed demonstrates this by showing the variation in energy level and the changes in network morphology during the life time of the agent and Wumpus whose behaviour is shown in Fig. 15. Figure 16 shows the variation in energy level of the agent and wumpus. The change in the energy level is a reflection of experiences. The increase in the energy level of the agent is an indication that the agent found the gold while the decreases indicated that the agent either faced a pit or was attacked by a wumpus. If the agent faces a pit then the decrease in energy is less compared to the decrease in energy due to the attack of wumpus. Here in this case as well, the energy of both agent and wumpus increases if they achieve their tasks. The energy level of

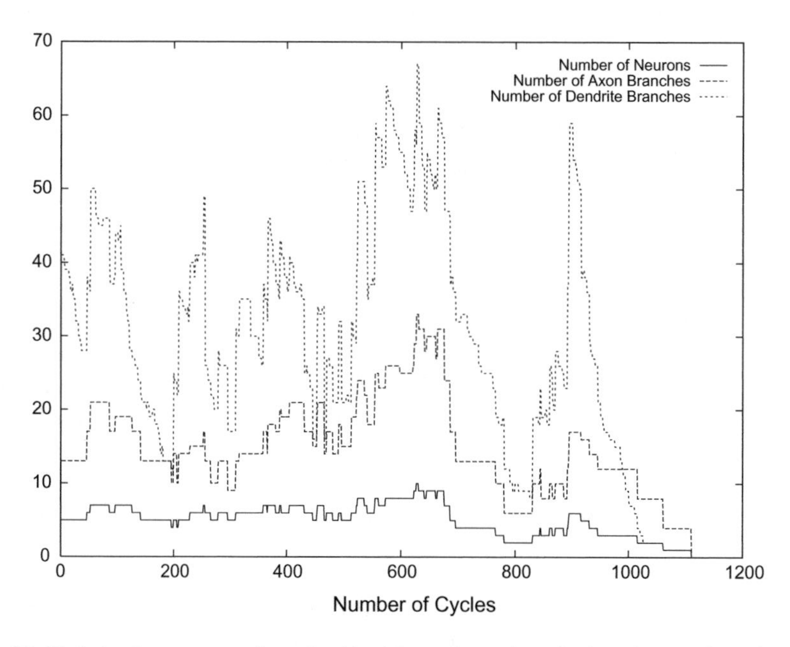

Fig. 17 Variation in neurons and neurites(dendrite and axon branches) against number of cycles, of the agent's CGPDN during the course of co-evolutionary task illustrated in Fig. 15. Variations in the network structure seems to coincide with the events illustrated in Fig. 16. Each step in Fig. 16 requires 5 cycles to complete

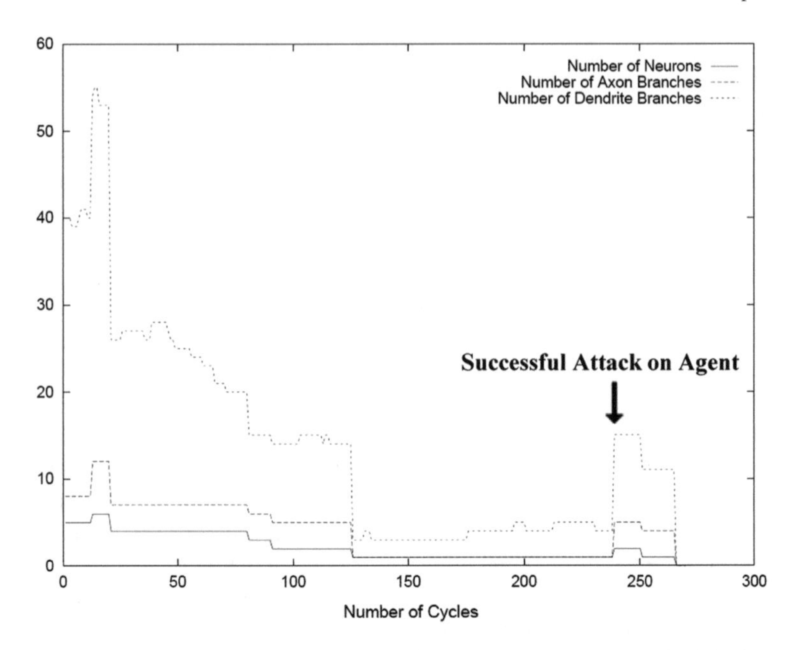

Fig. 18 Variation in neurons and neurites(dendrite branches and axon branches) during neural cycles of the wumpus CGPDN during course of the co-evolutionary task illustrated in Fig. 15. A continuous drop showing death of neuron and neurites causing death of wumpus at cycle 270 (step 54 in Fig. 16)

the agent drops by 60% if the wumpus attacks it. Strangely, all the neurons in the wumpus CGPDN network die shortly after the attack. Figure 17 shows the variation in number of neurons, axon branches and dendrite branches during the agent's life. Figure 18 shows the above mentioned numbers for a wumpus.

One of the observations is that mostly when the agent is caught by the wumpus, it is never able to get the gold again during its lifetime, since the interaction with the wumpus affects its network by causing the death of neurons. The work in this book tests the validity of this argument by increasing the initial energy level as indicated by Fig. 15. Through this, it is shown that the agent is able to get the gold even after being caught by the wumpus. But in such cases, the wumpus dies quickly after the encounter with the agent. This is due to the fact that the wumpus was about to die right before it catches the agent as shown in Fig. 15. The encounter seems to have increased its life span by triggering a brief increase in the number of neurons and branches.

The work also evolves the agent and wumpus with various values of initial energy. The influence of deleterious and beneficial environmental encounters is diminished by increasing the energy level to 300. As soon as the energy level is decreased to 50, the deleterious encounter outweighs the beneficial effects. The work in this book recommends an initial life of 100 to be a suitable balance. CGPDN builds a network and a learned behavioural memory in the environment. But this environmental responsiveness comes at the cost of possible dwindling away of all the neurons, since the networks are time dependent.

This chapter described a method of evolution of the CGPDN network on a classic AI problem of the wumpus world and also a co-evolutionary version in the same environment. Though the wumpus world problem has been modified, the results presented are quite promising. It was demonstrated that the system can solve dynamic task environments. Even if the network is tested on different wumpus world problems, the stability of the network is preserved. The agents show the ability to build their own environment map which is constructed during the development as a result of environmental experiences (Spector 1996; Spector and Luke 1996).

The next chapter will explore the learning abilities of the CGPDN in the game environment of checkers. It will also be demonstrated that how the two random networks develop a neural structure through evolution which can play checkers at a reasonable level of game.

Chapter 7
Checkers

The learning capabilities of CGPDN are explored to 'recognize' and 'learn to play' the arcade board game. The game of Checkers is selected as the arcade game since it was reconnoitred previously for learning by a number of AI algorithms. Like Wumpus world, checkers is also a grid based game however it is much more challenging and complicated compared to Wumpus world. Checkers is of great importance in the history of Artificial Intelligence and can be used as a test bed for evaluating the learning techniques (Dimand and Dimand 1996). The game of checkers is used here for demonstrating the capability of evolved networks to improve their ability to learn (level of play) by continuously playing against better opponents.

Building computer programs which can play games has been given importance ever since the rise of the Artificial Intelligence (Shannon 1950; Samuel 1959; Neumann 1928). Over the years various ideas have been presented about computer programs which can play games and there have been different checker playing programs. Modern understanding about the game of checkers is that this problem has been solved (Schaeffer and Herik 2002). The current world champion at checkers is a computer program called Chinook which is mostly based on a linear handcrafted evaluation function (Schaeffer 1996). The function considers the features of the game board such as piece count, king's count, trapped count, turn, run away checkers and other minor factors. The program can access the library of opening moves from games played by grand masters. It also has the complete endgame database for all boards with eight or fewer pieces. The program is based on human knowledge and there is no machine learning methods used in its development.

In spite of the fact that there have been many effective and outstanding methods used for the computer games, still it does not bear any resemblance to the human being's approach for games; since human beings do not use a numerical board evaluation function and they do not employ minimax. Human beings learn the game through experience. The model discussed in this book is based on the ability of learning which can be encoded in the genotype. After execution, it gives rise to a CGPDN network which can play the game efficiently.

G.M. Khan, *Evolution of Artificial Neural Development*, Studies in Computational Intelligence 725, https://doi.org/10.1007/978-3-319-67466-7_7

1 Checkers: The Game

The checkers board is available in different dimensions throughout the world. The Standard English checker's board is 8×8 with alternating light and dark squares. There are two players in the game of checkers which sit opposite to each other. Every player has 12 pieces. The pieces are of two different colours which are used to differentiate between the pieces of the two players. At the start, these pieces are placed on alternating squares of the same colour which are closest to the player's side. Usually the player with the dark coloured pieces starts the game. The pieces can only move diagonally only one square at a time, unless there is possibility of jump. Jump can take place when a piece moves diagonally over an opposing piece and land in an unoccupied square on the other side. If a piece jumps over the piece of opponent, the opponent's piece is removed. Multiple jumps can also occur if there is a jump possible after the first jump. If a piece reaches the first row of the opponent, it will become a king. Although a normal piece can move diagonally forward, the king can move diagonally in both directions. Jump has the highest priority among other moves. If there are multiple jumps available, then it depends upon the user's choice to make the move. However, a jump with king has higher priority. When a player has no pieces left or cannot make a move; he is considered the loser.

1.1 Experimental Setup

The model is a CGP Developmental Network (CGPDN). For the sake of checkers game, the model is arranged as under:

(1) The CGPDN of every player has neurons and branches which are located in a 4×4 toroidal grid.
(2) The initial number of neurons is 5, with each neuron having a maximum number of dendrites equal to 5.
(3) The maximum number of dendrite branches are chosen to be 200, and the maximum number of axon branches 100.
(4) The maximum statefactor of the branch is 7 while that of the soma is 3.
(5) The mutation rate is 5%.
(6) The maximum number of nodes per chromosomes is 200.
(7) The maximum number of moves that a player can make is 20.
(8) The chromosome length is 800 integers.

The signal and other parameters are represented by a 32-bit integer. The nodes operation in CGP genetic code are 32- bit logical operations.

1.2 Fitness Calculation

Both the agents can make only a limited number of moves and the fitness of the agents is accumulated at the end using the following equation:

$$Fitness = A + 200N_K + 100N_M - 200N_{OK} - 100N_{OM} + N_{MOV}, \text{ Where;}$$

N_K is the number of kings
N_M is the number of men of the current player
N_{OK}, N_{OM} are the number of kings and men of the opposing player
N_{MOV} is the total number of moves played.

The value of 'A' is 1000 for a win while it is 0 for a draw. A limit is set on the maximum number of moves as it reduces the computational time in assessing the abilities of poor game playing agents. In case the number of maximum moves allowed is reached even before any player wins the game, then the value of 'A' will be '0' which means that it will be considered a draw. The number of pieces and the type of pieces is then used for assessing the fitness value of the agent. The number of kings is multiplied with twice the number as that of the normal pieces, for conveying the importance of the kings. These numbers are arbitrarily chosen but can be modified by the users. The number of moves is of significance since it encourages the agents to make more moves and live longer.

1.3 Inputs and Outputs of the System

The input to the system is in form of 32 element array of board values. Each element represents a playable board square. These 32 inputs can represent five different values depending on the square of the board which are represented by 'I'. The values of 'I' are as follows:

- $I = 0$ for an empty square.
- $I =$ Maximum value $(M) = 2^{32} - 1$
- $I = (3/4)M$ for a piece.
- $I = (1/2)M$ for the opposition piece.
- $I = (1/4)M$ for opposition king.

These parameters are chosen intuitively and can change.

The board inputs are applied in the form of pairs to all the sixteen locations of the CGPDN grid. The playable squares are 32 as shown in Fig. 1. Figure 1 shows the interfacing of the checker board with CGPDN. There are input axo-synapse branches for every playable board positions. The inputs can run the axo-synapse electrical CGP to provide input to the CGPDN.

The input potentials of the two board positions and the neighbouring dendrites branches are applied to the axo-synapse chromosomes. This chromosome produces

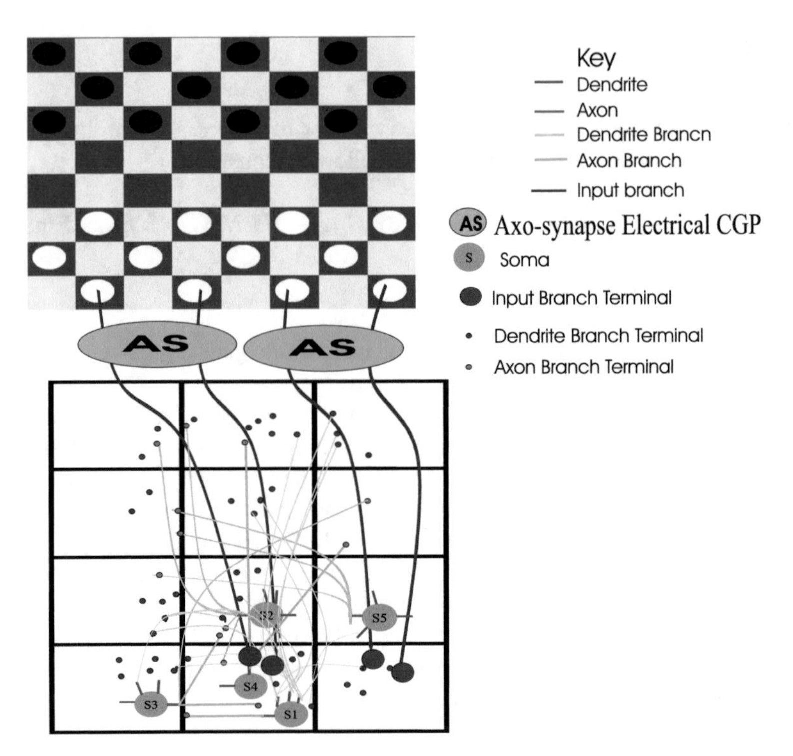

Fig. 1 Interface of CGPDN with Checker board. Four board positions shown being interfaced with the CGPDN

the updated values of the dendrite branches in that particular CGPDN grid square. There are two branches for two board positions in the CGPDN grid square. After this, the axo-synapse chromosome is run for every square individually starting from square one and finishing at square sixteenth. Output has two forms. The first form is used to select the piece to move while the second to decide where the piece should be moved. There is an output dendrite branch in the CGPDN for every piece on the board. Every piece has a unique ID that represents the CGPDN grid square in which its branch is located. These branches have a potential which is updated during the CGPDN processing. The possibility of moving the piece is dependent upon the values of the potential. The piece with the highest potential is moved, however if there are pieces which can jump; then the piece with the highest potential among them is moved. The king which can jump has the highest priority. In case there are two kings who can jump, then the king with highest potential jumps. There are also five output dendrite branches which are present in various random locations in the CGPDN grid. The direction of movement of the piece is determined by the average value of dendrite branches potentials. If any piece is removed due to a jump, its dendrite branch is removed from the CGPDN grid.

2 Co-evolution of Two Agents Playing Checkers

The method of evolving two or more systems together such that they affect each other's evolution is known as Co-evolution. These techniques have been applied to various games and have been found effective (Irving and Uiterwijk 2000; Wee-Chong and Yew-Jin 2003; Fogel 2002).

Competitive co-evolution using CGPDN is implemented where each agent compete with one another in the game of checkers. There are 5 genotypes in the agent's population. Each of the first five agent population members are tested against the best performing second agent genotype of the previous generation in a single game (Stanley and Miikkulainen 2004). The co-evolution happens as under:

- Both agents are provided with a population of five genotypes
- In every generation, the genotypes of the first agent are tested against the best genotype of the second agent and vice versa.
- The fitness of all the genotypes is evaluated and the agent with the highest fitness is selected as the parent for the upcoming generation. Both the agents are then mutated for producing four offspring (1+4 Evolutionary Strategy). The arrangement is made in such a manner that the fifth one is the parent and the first four are offspring for the next generation. So the fifth one is the optimum performing genotype from the last generation of both agents.
- The process mentioned above is repeated until a solution is found or maximum number of generations is attained.

During the first generation, the initial random CGPDN structure of both the agents is same. Initially at the start of the experiment, the input to the agent playing black is applied with the board values. Axo synapses are used to apply inputs to its CGPDN. The CGPDN network is then run for five cycles, during which the potentials of the output dendrite branches are updated. The average of dendrite potentials is taken and utilized to decide the direction of movement for the corresponding piece. The CGPDN also updates the potentials of the output branches for the pieces. The updated potential values are used for deciding which piece is to be moved. In case if there is any jump available, then the jump has the priority. If there are multiple jumps available, then the one with the highest potential has the priority. This process is also repeated for the opponent and it continues repeating until the game ends.

The game will stop if any of the following things happen

- The CGPDN of the agent or its opponent dies.
- The opponent loses all its players.
- The agent or opponent cannot move anymore.
- The allotted number of moves is taken.

2.1 Learning 'How to Play'

Checkers is a difficult game. In order to learn the game, the two agents start playing with only a few neurons which have a number of dendrites and branches. The agents build a developmental network which is capable of solving the task while keeping a stable network. They also find a way for processing the environment signals and differentiating among them. They understand the spatial layout of the board. The agent then develops a memory or knowledge about the meanings of the signals from the board. They also develop the memory of previous moves and determine whether they were beneficial or deleterious. The agents also need to know about the benefits of making a king or jumping over. The agents perform all these mentioned tasks while playing the game. Agents have the ability to learn from each other about favourable moves as the generations pass by. The learning is transferred from one generation to the other through genes.

In order to find out about the performance of the more evolved agents, the work here presents an example in which a more evolved agent is tested against a lesser evolved agent. This verifies that the agents of later generations perform better. The lesser evolved agent will almost always lose at the hands of the more evolved ones, however sometimes there can be a tie between the two agents. Even in case of tie, the more evolved agent will end up having more kings and pieces compared to the lesser evolved.

The variation in the fitness of a more evolved agent against the fitness of a series of agents from earlier generations (less evolved ones) in the same evolutionary run has been presented in Fig. 2. It also explains the cumulative fitness graph, showing the summation of fitness over various generations. It can be clearly seen that the more evolved agent playing white has higher fitness values compared to the less evolved agent playing black. The behaviour of the more evolved agent is not uniform as in one run; the more evolved player can beat its less evolved opponent (opponent at generation 100) by a huge margin; while in another run the might not play well against the less evolved player of generation 100. Although it might do very well against the less evolved player at 150th generation.

In Fig. 2 it can be seen that initially the agent beats the opponent by a margin greater than 4 kings or 8 pieces, within twenty moves. This difference reduces with the decrease in evolutionary training differences. Figure 4 showcases a more evolved agent (in white) playing a lesser evolved agent (in black from generation 5). When the game starts, the initial board position is fed into the CGPDN of the agent (black); the CGPDN then runs for five cycles. After this the CGPDN makes decisions about which piece is to be moved and where it should be moved. Table 1 shows the sequence of moves up to move 31 (Fig. 3).

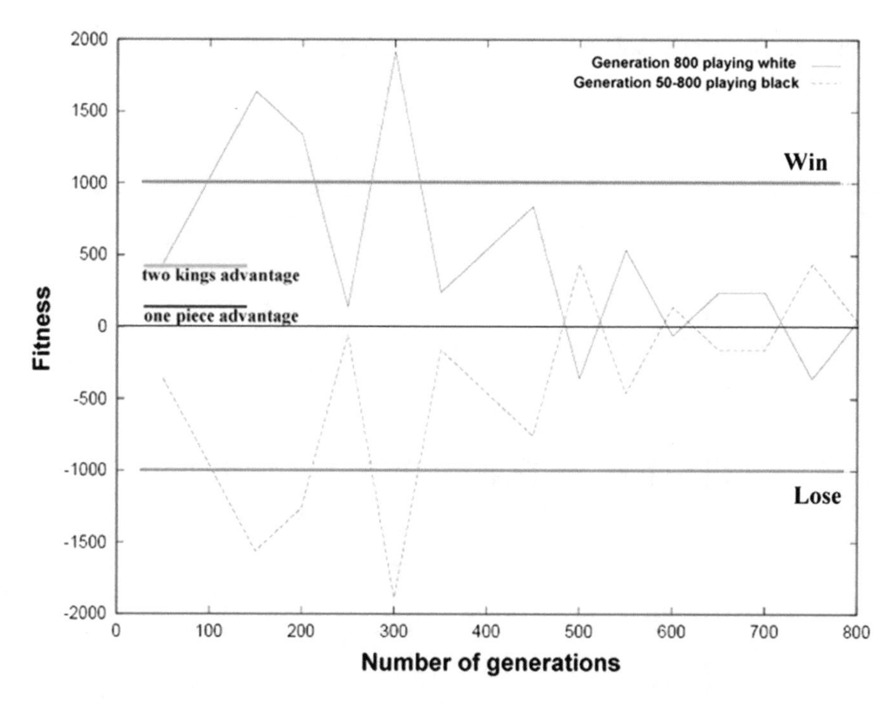

Fig. 2 Fitness variation of a more evolved agent playing checkers against different generations of the lesser evolved player

The game goes fine up till the 8th move when the white takes a black piece. Then at the 10th move, black has the opportunity to take a white piece in two ways. After an ensuing series of forced exchanges, white pieces move further up the board compared to black. This is a sensible move as it is the only way one can obtain kings. Move 19 is a catastrophic one for the black as white can take two pieces and acquires a king. Move 24 is also a disastrous one for black pieces as it loses two more pieces to the white pieces. Both the players develop the ability to protect their pieces by placing pieces behind them. The defensive move occurred at 4, 7, 15 and 16. There were 10 possible moves at move 4 when there were five pieces which each could move in two directions. However the player selected only one for defending its previous move. At move 7, there were 8 possible moves. The player again defended its previous move. At move 15 there were 8 possible moves however the player selected the move to defend its piece. Finally at move 16 there were 12 possible moves, however the player defends its piece again. In move 31, the black piece moves to the edge which is interesting as edges are safe places to be in. The white wins the match in

Table 1 The first 31 moves of a game between a more evolved player (white) against a less evolved player (black)

Move number	Move	Comment	Move	Comment
1	B1 10–13	Opening		
2			W2 23–19	
3	B3 11–15			
4			W4 28–23	Defend
5	B5 5–10			
6			W6 23–20	
7	B7 1–5	Defend but . . .		
8			W8 20–11	Takes
9	B9 6–15	Takes		
10			W10 22–18	Offer
11	B11 15–22	Takes		
12			W12 26–19	Takes
13	B13 13–22	Takes		
14			W14 27–18	Takes
15	B15 2–6	Defend		
16			W16 30–26	Defend
17	B17 6–11			
18			W18 32–28	
19	B19 9–13	Blunder		
20			W20 18–9	Takes
21			W21 9–2	Takes, gets K
22	B22 11–14			
23			W23 28–23	
24	B24 14–18	Blunder		
25			W25 21–14	Takes
26			W26 14–5	Takes
27	B27 7–11			
28			W28 31–27	
29	B29 11–14	Blunder		
30			W30 19–10	Takes
31	B31 12–16	Move to edge		

48 moves with one king and eight pieces left. Both white and black players play the game sensibly, however the black made some blunders; and white made some good moves which resulted in a victory for the white. This example demonstrated the learning ability of the agent and showed how the two developmental programs learned to play checkers through co-evolution.

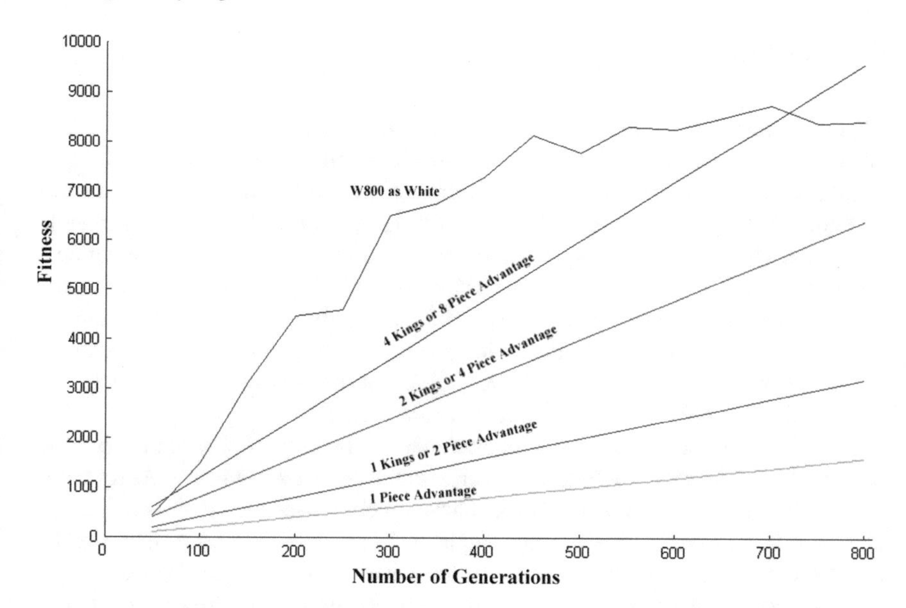

Fig. 3 Accumulated fitness variation of more evolved agent playing checkers against different generations of lesser evolved player

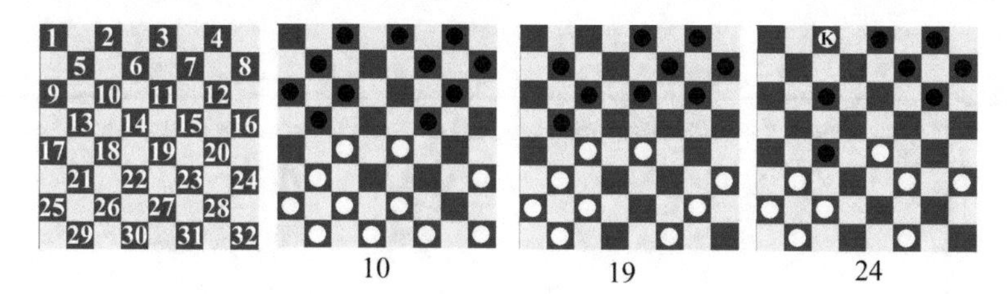

Fig. 4 Labelled Board and positions at different stages of the game. Numbers beneath boards show the board at moves 10, 19 and 24

3 An Agent Plays Against a Minimax

We also present CGPDN checker players that are investigated for its learning behaviour against a minimax based checkers program (MCP) instead of using co-evolution. MCP program plays at a higher level. Every genotype of the agent plays five games against the MCP. The agent starts the game from a random network. The best playing genotype based on fitness is selected as the parent for the new populations and is promoted to the next generation along with four offspring without any change.

In this case, the MCP makes the first move and the updated board is then applied to the agent CGPDN. The CGPDN network is then run for deciding the piece and the move. This process is repeated until the game is stopped.

3.1 Results and Analysis

MCP always had the upper hand over the agent, so it is very tough to find out the variation in fitness of agent during the course of evolution whether it has learning or not. Despite learning a lot of moves during the course of the game, the agent is still not able to beat the MCP. That is why the agent's fitness level is low and varies randomly. The MCP produces a database of game, which it uses for computation of the next move. Due to this reason, the MCP plays a different game every time even against the same opponent. This makes it difficult for the evolution to select the best genotype from the next generation. The genotype of the best agent is promoted to the next generation; however it produces different fitness values. That is why it is quite difficult to obtain any improvement against MCP.

In order to check if there is any improvement in the level of play of the agent, we have evaluated the more evolved agents against the lesser evolved ones. As discussed earlier, the more evolved agents always have an edge over the lesser evolved ones showing the improved level of play. Figure 5 shows the fitness of a more evolved agent against a series of agents from the earlier generations. It can be clearly seen that the more evolved agent (white) always beats the less evolved agent (black). Figure 6 shows that the more evolved agent beats the early ancestors by a huge margin, however with the progress in evolution; this margin reduces. This shows the

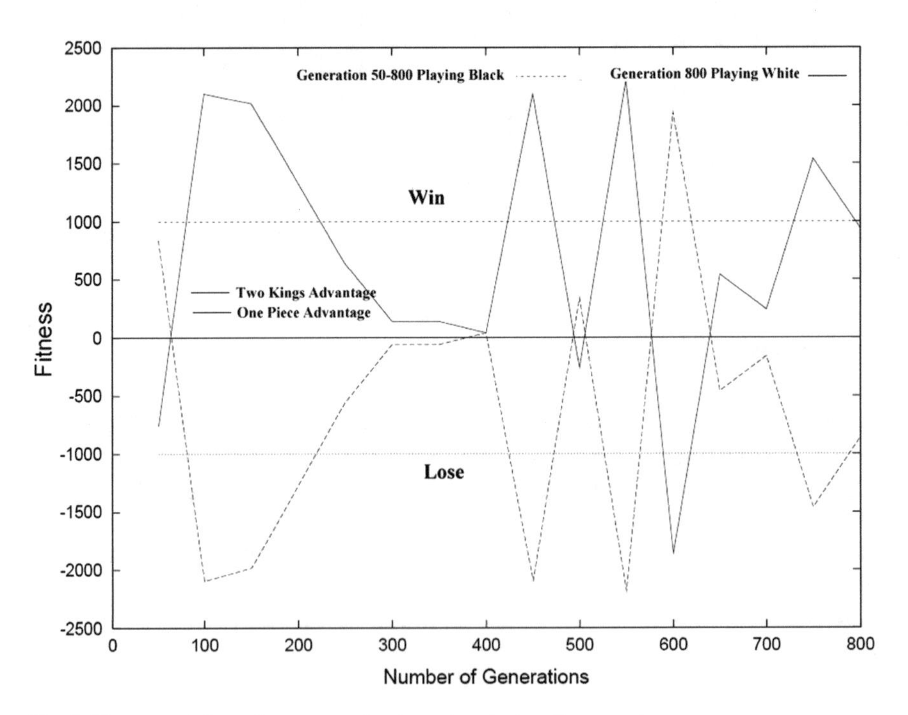

Fig. 5 Fitness variation of a more evolved agent against earlier ancestors

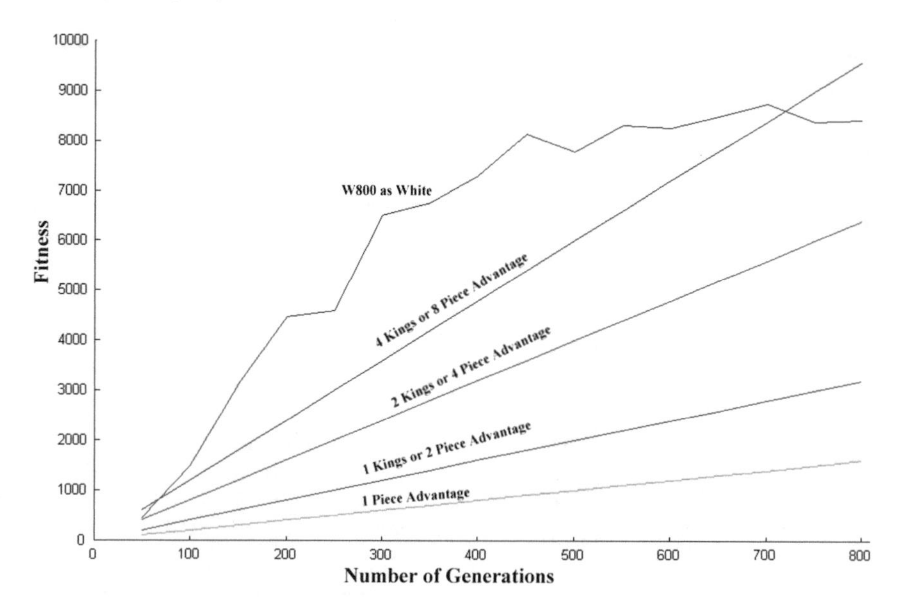

Fig. 6 Accumulated fitness variation of a more evolved agent against ancestors

increase in learning ability of agents over the course of evolution. Table 2 presents the game between a more evolved agent and a lesser evolved agent. Figure 8 shows some of the important board positions during the game scenario. The more evolved agent plays black while the ancestor agent plays white now.

Initially the board position is fed into the CGPDN of the agent playing black; the CGPDN is run for five cycles. Then the CGPDN makes the decision about which piece to move and where to move. The first 4 moves that the players make are normal moves, however, the white pieces move further advanced. The black leaves an empty square on its back after move 4. This move results in a catastrophe with move from the square 12 to 15. The white piece takes two black pieces and becomes a king. Then some reasonable moves are made. Again the black piece moves from square 10 to 14 allowing the white king to take two black pieces. Despite black taking some white pieces in move 16, white proves to be too strong to be handled and wins the game.

This example just described the difference in the level of play of the two agents.

A few more examples ascertain the learning ability of the agent. Experiments demonstrated that it is very tough for an agent to learn from a highly skilled system. Figure 7 shows that it is hard to asses if any learning can take place in case an evolved agent plays against the MCP. Figure 9 shows the changes in the fitness of an evolved agent from the 2000th generation (white pieces) while playing against lesser evolved agents from earlier generations. The plot shows the average fitness in consecutive games of the two agents. It is evident that the more evolved agents hold an upper

Table 2 Game played between a more evolved player (white) against a less evolved player (black)

Move number	Black	White
1	B1 10–14	W1 24–20
2	B2 6–10	W2 23–19
3	B3 14–23	W3 28–19
4	B4 3–6	W4 27–23
5	B5 12–15	W5 19–12 12–3
6	B6 11–14	W6 23–19
7	B7 14–23	W7 31–28
8	B8 10–14	W8 3–10 10–19
9	B9 4–7	W9 19–14
10	B10 7–11	W10 14–7
11	B11 23–27	W11 30–23
12	B12 8–12	W12 7–16
13	B13 9–13	W13 23–19
14	B14 2–6	W14 21–17
15	B15 6–11	W15 17–10
16	B16 5–14 14–23	W16 28–19
17	B17 11–14	W17 19–10
18	B18 1–5	W18 10–1

hand over the lesser evolved agents. The network of both agents, develops in the game series, so they start the new game with the developed architecture.

Figure 9 shows that the well evolved player does not play any better against the later generations compared to the earlier generations. The MCP updates its database after every game. Its approach in every game is very different even if it is playing the same opponent. This indicates that it is tough for the agent to maintain its previously gained higher value of fitness in the following game. For achieving higher fitness, the agent has to alter the way it plays a game. If the less evolved players do not change the way they play a game (it happens when they are playing close ancestral relative) they are easily beaten by the well evolved players, however they are hard to beat if they change the way they play a game (it happens when they are playing distant ancestral relative).

In another example, an agent begins every game with a random network while the other agent is allowed to retain its network and develop it over the course of the games. A well evolved agent of 2000th (white pieces) generation competes against an agent from 50th generation (black pieces). The agreed rules for the games are that both agents can make 20 moves and the 50th generation agent starts playing every game with the same initial random network on which it was initially trained;

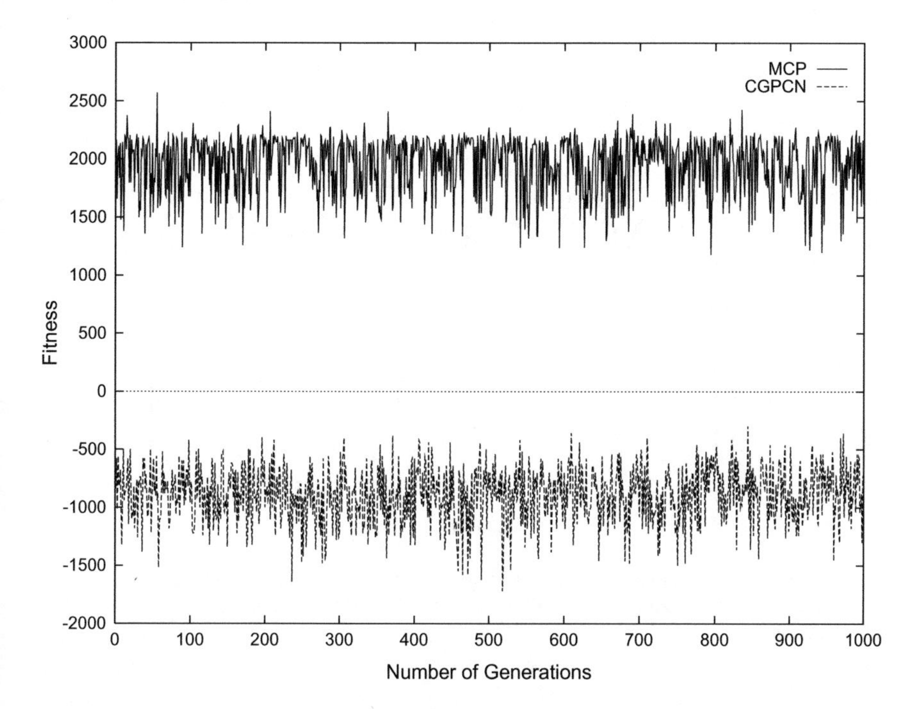

Fig. 7 Variation in fitness of CGPDN against MCP while playing checkers during course of evolution

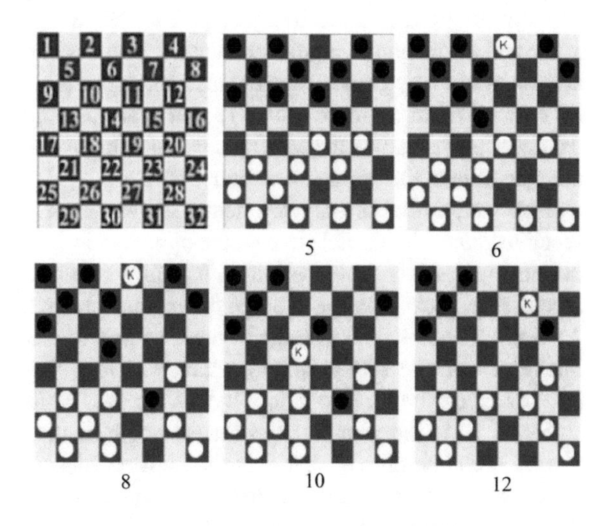

Fig. 8 Labelled Board and positions at different stages of the game. Numbers beneath boards show the board at moves 5, 6, 8, 10 and 12

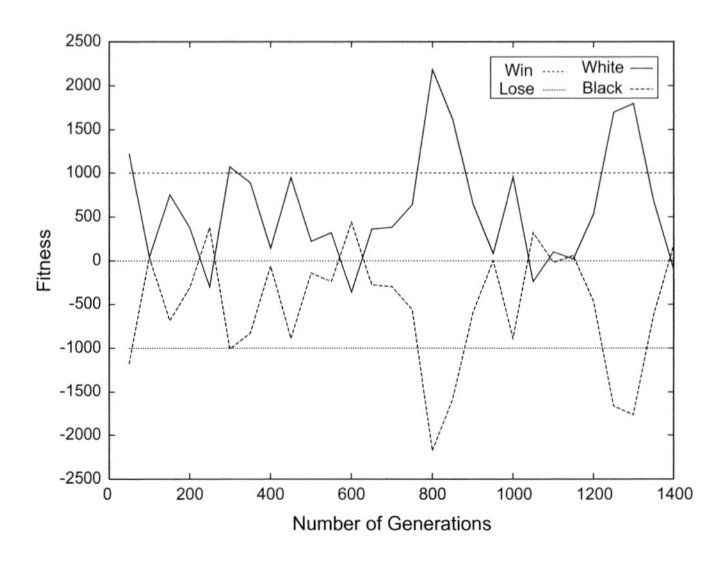

Fig. 9 Graph showing the fitness (Averaged after every five games) variation of a well evolved agent (white) playing checker against different generations of less evolved player (Black) playing five games each

while the 2000th generation agent continues with the network it had at the end of the previous game. The genotypes do not change; only architecture of CGPDN and its morphology is allowed to develop. The fitness of both the agents are calculated at the end of the game. Figure 10 shows the fitness variations of the agents obtained at the end of every game. They are also some peaks at various stages which indicate the case in which the more evolved agent beats the opponent within 20 moves. When the agent plays against the MCP or any other highly skilled checker program, it was not able to beat the opponent within 20 moves.

However, during the developmental stages; when the agent plays 5 or more consecutive games, it was able to beat the opponent in 20 moves. The agent continues to develop and play without evolution, its ability to beat the opponent within 20 moves increases as proven by the average fitness stay above the x-axis in Fig. 11.

Figure 11 shows the averaged fitness of five consecutive games. The un-dashed line represents the 2000th fitness, which is always above zero. This indicates that its performance is better throughout the 500 games, while it is developing its network.

Figure 12 is the accumulated fitness graph of the well evolved agent over 500 games. It can be clearly seen that the network, despite the changes in the network during every game, is able to sustain its integrity of getting higher fitness over the less evolved agent. This shows that the agent does not forget playing checker better.

Figure 13 shows the changes in the number of neurons and neurites of a well evolved agent of 2000th generation during the games. The figure clearly shows that the network initially varies a lot. At some point, it reduces to the minimum structure. Later on it stabilizes to structure with a fixed number of neurons and neurites.

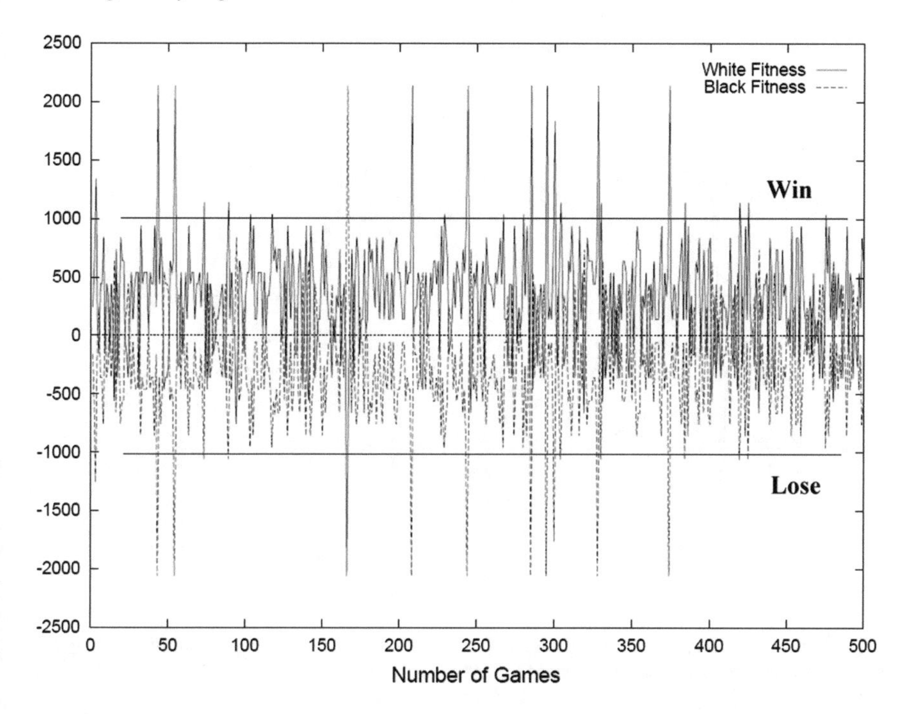

Fig. 10 Graph showing the fitness variation of a well evolved agent (white) against a less evolved agent (black)

This is very interesting as the network can still develop, though the number of neurons and neurites become stable; the branches are still able to migrate and weight of neurites being updated. The network is not trained to find a small network, however in consecutive games; it continues to change until it finds a minimal suitable structure which can play better as evident from the accumulated fitness graph in Fig. 12. Deep analysis of the examples can show that even the updated network repeats its initial moves, which results in two double jumps and the opponent losing its 4 pieces. This shows that opponent always starts with the same initial condition and repeats the same initial moves. In the first 100 games, it is observed that the agent repeats its initial 8 moves almost every time which causes the agent taking 4 pieces. This is because the agent does not know when one game ends and the other starts. It starts the other game with a different network which forces the opponent to repeat the same mistakes. As a result of the repetitive mistakes, the opponent loses the game. This indicates that the agent can respond to the variations in the board positions and is able to make the same moves with a different network. This is the indication of a stable behaviour even when the obtained CGPDN is changing.

The discussed examples clearly demonstrate that the system can improve with evolution. The CGPDN network can develop during the game and it is the cause for the intelligent behaviour of the agent. The agent can learn experimentally when it is trained in a consecutive game scenario. This also causes the development of a net-

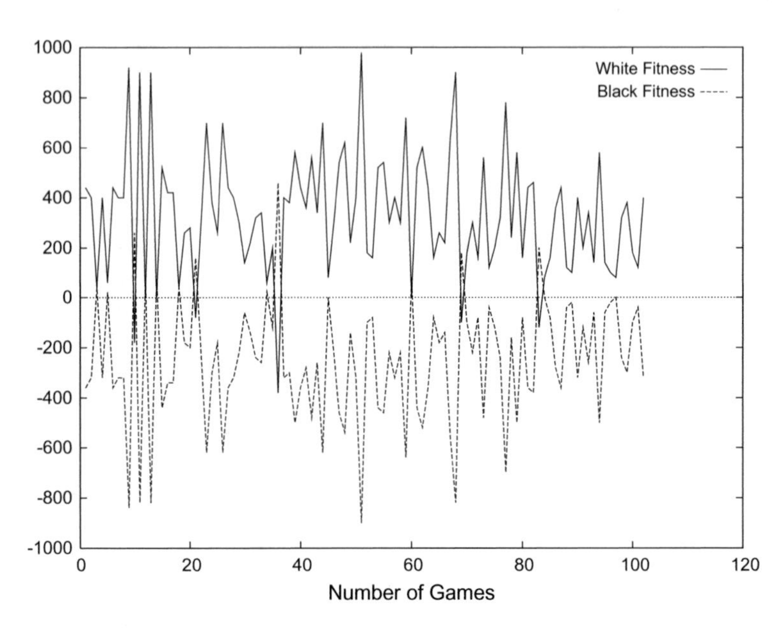

Fig. 11 Graph showing the average fitness variation of a well evolved agent (white) playing five games of checkers against a less evolved agent (black)

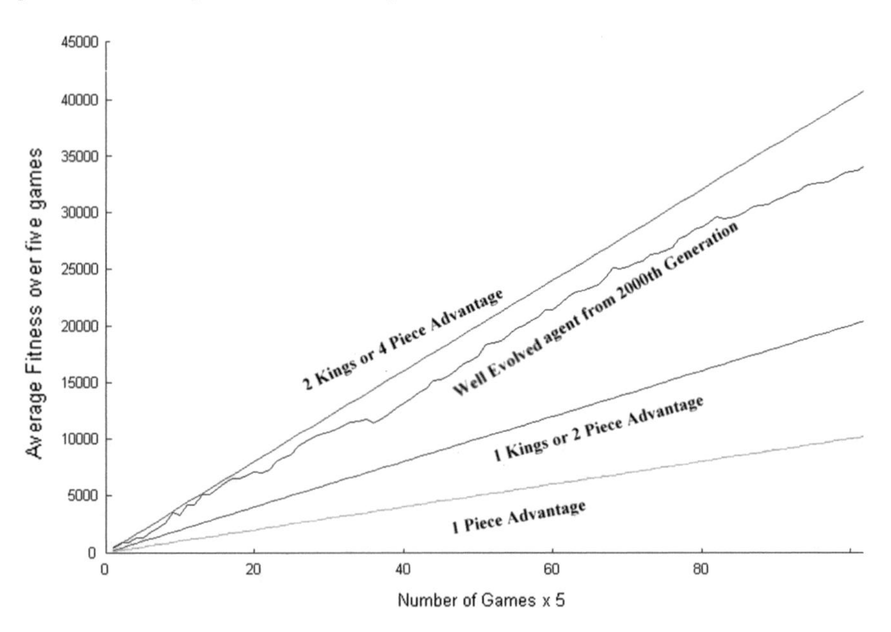

Fig. 12 Graph showing the accumulated fitness of a well evolved agent (white) playing checker against a less evolved player

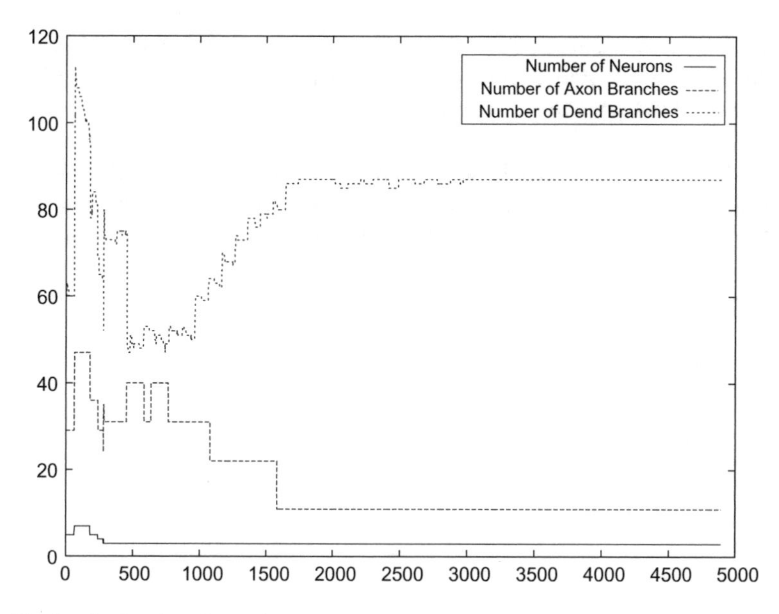

Fig. 13 Graph showing the variation in number of neurons and neurites of a well evolved agent(white) over five hundred games

work which can sustain its architecture and continue performing better with different architectures. The examples also show that the CGPDN network can maintain the consistency in winning the games, even if the architecture is updated. The CGPDN network also obtains a default minimum structure which is suitable for it. The examples show that the network can repeat similar moves, every time a new game is started. This indicates that the network has the ability to carry out intelligent action repeatedly in the board environment, even if the network is varying. The examples reveal that the CGPDN can adapt to varying environment, sustain its architecture and it can solve tasks even while developing. The robustness and adaptation of the system makes is suitable for different application in the field of science and technology.

The network presented is a biologically plausible developmental network which has a dynamic morphology. It is able to handle the arbitrary number of inputs and outputs. The response to the following statement was presented up to some extent through the model demonstrated:

> Implementation of an autonomous computation system inspired by neuroscience which is capable of continuously learning and adapting in complex environments. A possibility or a dream

It was demonstrated that the CGPDN has the ability to adapt to the variations in the environment and to sustain its architecture which proves its robustness. The CGPDN system has the potential to solve tasks while continuing to develop.

However there is still a lot of work to be done in the CGPDN. Improvements can be made to produce intelligent behaviours which are much more like the mammalian brain. Implementing the CGPDN in hardware form will also be something to look forward to in the near future. This will be a major breakthrough if achieved as it will be opening the doors to heights never attained before in the field of science and technology.

Chapter 8
Concluding Remarks and Future Directions

Developing an intelligent network capable of learning and adapting in a complex environment without human intervention has been the main focus of this work. We will discuss to what extent the goals of this research have been accomplished and how significant are its contributions to the field of artificial intelligence.

At the beginning of this work, the following hypothesis was stated:

> Is it possible to implement an autonomous computational system inspired by neuroscience capable of continuously learning and adapting in complex environments.

From the evidence provided in Chaps. 6 and 7 of this book, it can be concluded that the hypothesis has been supported to a certain extent. The CGPDN introduced, is a biologically plausible computational network as explained in Chap. 5. The network produced has a dynamic morphology and is capable of handling an arbitrary number of inputs and outputs as explained in Chap. 5. The system is inspired by the biological brain. Chapter 5 introduces CGPDN, which incorporated ideas from the biological brain and neural development and allows the network to develop its architecture. The genotype of the network is evolved so that it develops an environment to produce a network that can solve the task.

The structure and operation of the CGPDN Model is inspired by principles of neuroscience, its general layout, operation and implementation is discussed in Chap. 5. The Cartesian Genetic Program (CGP), which is used as a genotype for the model and the method of interfacing CGPDN with external environments is provided in Chap. 5. Detailed explanations of the information processing in the CGPDN and neuron model have been discussed.

We have examined the characteristics and performance of the model in the context of an intelligent agent trying to solve a learning task called Wumpus world. Results are very promising and indicate that this system is capable of developing an ability to learn continuously within a task environment. We found that a network tested on a different Wumpus world preserves the sustainability of the network and develops the learning ability of the agent to avoid pits and the Wumpus.

G.M. Khan, *Evolution of Artificial Neural Development*, Studies in Computational Intelligence 725, https://doi.org/10.1007/978-3-319-67466-7_8

The network develops while solving Wumpus world in a dynamic manner in search of desired neural architecture capable of solving the task. Different experiments are performed with CGPDN having different initial populations and Wumpus world environments to test whether the agent is capable of solving the task.

We have also tested the capability of CGPDN in a co-evolutionary competitive learning environment. Two antagonistic agents are provided with different CGPDN which grow and change in response to behaviour, interactions with each other and the environment.

From the results in Chap. 6 we found that the agents can learn from their experiences and interaction with each other and appear to build a map of their environment, providing evidence of learning and memory development.

We have tested CGPDN in the context of playing the well-known game of checkers. We co-evolve two agents having CGPDN as their main processing unit. After a number of generations we tested the well evolved agents against the lesser evolved ones and the results showed that the well evolved always beat the less evolved. Thus evolution had improved the level of play of agents. We also evolve the agent against a Mini-max based checker program (MCP) and evaluated more evolved programs against the lesser ones, once again we obtained similar results, with more evolved players always beating the lesser evolved players.

Chapter 7 provides a detailed description of the game of checkers, and demonstrated how two random networks end up developing a neural structure through evolution, that can play checkers at a reasonable level.

Our results are encouraging and demonstrate that the system improves with evolution. The CGPDN morphology continues to change during a game, allowing the agents to adapt learning behaviour.

We trained the agent for experiential learning, by allowing the agent to play five games each against MCP without changing its network and allowing it to develop. The agents always start with a random network in the first game and inherit its updated network from the previous game in the subsequent games. This allows the agent to have experience of more than one game scenario. Again we have evaluated more evolved agents against lesser ones in five game scenarios, with more evolved agents always beating the lesser ones as evident from the results in Chap. 7.

In order to test whether CGPDN maintains its integrity of solving the task even when its network is developed into a completely different network rather than the one it is trained on, we allowed a more evolved agent to play five hundred games against a lesser evolved agent so that the well evolved agent can inherit its network from previous games and continue to develop, whereas less evolved agents always started from a random initial neural structure. The experiment produced interesting results with the more evolved agent continuously beating the less evolved agent, even when the network is changed a lot during development, and stabilizes at a minimum structure at the end. Additionally, close observation of the games, reveals that the agent repeats similar initial moves causing it to get an initial advantage over the opponent, although its network is different (more developed) from the one in the previous game. This shows that the network has the capability of carrying out the same intelligent and useful behaviour even though its network changes. This further suggests that the network may not suffer from catastrophic forgetting.

In future we are planning to train the network in co-evolutionary environment for experiential learning. At the moment the system resources take long to evolve, because fitness evaluation is very slow with limited computational resources. We are planning to implement the network on high speed hardware, to speed up the evolutionary process.

We studied different models of ANNs and neuroscience literature and implemented a system at neuron level, and provided it with an artificial environment where neurons can interact with each other. The basic neuron model we produced is based on biological studies of neurons, their development and signal processing mechanisms. These neurons can grow and produce complex neural structures based on the requirements of the task.

Our model is inspired by the work on the neurodevelopment techniques; we evolve the rules for development of the neural architecture and their internal processing.

Chapter 5 has introduced a novel technique called CGPDN, which is a developmental model of neural computation using CGP inspired by the biological brain. This network is capable of self-development during the task environment. The network is produced using biological principles of neurons and neural computation. We allow the genes of the network to evolve, so that it develops neural structure capable of learning.

Experimental evidence in Chaps. 6 and 7 has shown that the system might be developing a memory of its experiences during development. The research in this book provides the idea of next generation of neural networks by adding more biological plausibility to neural networks. In this model most of the processes inside the neuron are considered and a CGP model for these processes has been introduced. The input/output to the system is also applied through dendrite and axon branches.

The model also provides a new dimension to Cartesian Genetic Programming by using a collection of chromosomes. The group model of chromosomes in this case is inspired by biological neurons. Genetic programming is used to encode the gene information inside the neuron and to evolve intelligent behaviour in the neuron.

We used evolutionary strategies for selection process in evolving the system (CGPDN), and Cartesian genetic programming to find the unknown functions inside neurons for their development and electrical processing. CGP is used because it is simple and easy to implement. Conventional ANNs could in principle, also be used instead of CGP to find the unknown functions inside neurons. We tested the system (CGPDN) in a co-evolutionary task environment where intelligent behaviour was rewarded in wumpus world (explained in Chap. 6) and checkers (see Chap. 7). We used neural development techniques to develop the architecture of CGPDN in a task environment (post evolution).

Our research model is based on biological morphology of neurons and allows the neurons in our system to have random number of dendrites and branches that grow and branch to produce their own desired structure based on the functionality required. We divided the neuron into three main parts, the soma (cell body), dendrites with branches, and an axon with a number of branches. Inspired by leakage currents in neurons, we reduced the potential of all the branches and soma after every cycle giving it a sense of time and change.

We have implemented the mechanism of action potential generation in which the signals are received from all the neurons in the environment through dendrites, deciding on whether to fire an action potential or not.

We introduced the process of synaptic plasticity in our network by adding weights to branches and soma which is updated at runtime based on the biological synaptic plasticity.

One of the main problems with ANNs is that if their weights are changed they are unlikely to be able to solve the same task or give better performance. Whereas the CGPDN seems to retain its integrity of solving the task and improving its performance while its morphology and weights are changed during the task environment. The results obtained from the environments of Wumpus world and checkers suggest that this does not cause a problem for the CGPDN. The network when explored on different Wumpus worlds seems to have ability of keeping stable network and goal driven behaviour causing it to avoid pits, and wumpus and collect gold. CGPDN is a developmental network, capable of self-configuration during a task environment. It is capable of handling variation in the number of inputs and outputs at run time. The genotype of the network is evolved to get a generalized capability of learning and functionalities.

Based on the above results and analysis, we can conclude that CGPDN has the capability of adapting to changes in the environment, and of sustaining its architecture. It has a potential for solving a task while continuing to develop. This robustness and adaptation capability of the system makes it useful for a number of applications both in science and technology.

This model should provide a completely new area of research that we hope will be capable of producing intelligent behaviour that in some ways is more like the mammalian brain.

This research is of a fundamental nature. There are very few developmental approaches to neural networks, currently being considered and the evolved compartmental model of neural function is entirely novel. The potential benefits of this research are great. If it is made possible for a program to evolve via learning through experience then there are many potential applications. Our initial research on this topic has been highly encouraging.

In future work, we are looking forward to explore CGPDN in a more challenging and diverse environments to see if? it is possible to evolve a general capability for learning? We will also look into the dynamicity of CGPDN for its ability to solve problems faster and accurate through repeated experience of its task environment (online learning).

We will be looking for other applications in future work, as the network seems to have dynamic behaviour, and is capable of solving a range of dynamic problems.

In future we propose to investigate the current model of CGPDN for its capability of learning through experience. The genetic programs in CGPDN are encoded as digital circuits built from multiplexers. Such circuits fit well with modern mux-based Field Programmable Gate Array Architectures (FPGA). The idea is to develop the network on high speed hardware, in order to speed up the evolutionary process and test CGPDN in co-evolutionary environment for experiential learning. Also the network has many parameters and it is difficult to test the importance of different parameters and chromosomes, as the evolutionary processes take a long time on desktop machines (PCs).

We have the following two main objectives for future work:

- To analyse and refine the current model of Cartesian Genetic Programming Developmental Network (CGPDN) and show that it allows generalized learning through experience in the context of playing checkers
- To build a prototype CGPDN on electronic hardware

Objective 1: **Analysis and refinement of the CGPDN to demonstrate experiential learning**

Our objective is to analyse and refine the current model of CGPDN and train it to achieve a high standard of checkers by a combination of evolution and learning. We wish to evaluate and show that evolved genotypes when run in the task environment improve their performance through experience alone (i.e. post evolution). In addition we wish to understand the relevance and role of various parameters and predefined rules in the model with respect to evolvability and learning capability. There are various ways that an arbitrary number of inputs may be introduced into CGP chromosomes. We wish to evaluate the effectiveness of these and to understand and visualize morphological changes and the changes in activities of the CGPDN during long periods of training. The task will be divided into following four sub tasks:

(1) *Analysis and Refinement of CGPDN model*: The CGPDN is complex as it makes use of seven evolved circuits along with many predefined rules and parameters. The importance and role of these rules need to be understood in more detail. Investigations into various ways of applying an arbitrary number of inputs to CGP chromosomes and assessing their effectiveness.
(2) *Investigation of various parameters*: Investigation of importance of various fixed parameters in the CGPDN with regard to the speed of evolution and the degree of acquisition of its capability for learning.
(3) *Visualization and Statistical analysis of CGPDN*: Visualization and statistical analysis of morphological, 'electrical' activity, and life-cycle changes during the development of the CGPDN from its initial random state and from its 'mature' untrained state to its 'mature' trained state.
(4) *Investigation of fitness functions*: Assessment of various ways of accumulating fitness to encourage experiential learning, including repeated task environments. Introducing agent energy and mechanisms for accumulating fitness within task environment. Analysis and demonstration of experiential learning.

Objective 2: **Hardware implementation of CGPDN**

Hardware implementation can be considered the ultimate goal of a neural network approach, as it allows a direct interface with the real world and can perform parallel computation that can only be approximated in software. For an evolutionary implementation, hardware has an even greater importance, since it is only through the speedup it allows, that the complex computation required by genetic algorithms can be scaled up to large networks.

Since all the processes inside neurons in our model are based on simple multiplexer operations, implementing the evolved genotype in hardware will allow the neurons to do processing in parallel rather than serially. We expect a considerable increase in processing speed, as at present the program runs on a sequential computer that does multiplexer operations one at a time, whereas an FPGA implementation would allow the multiplexers to operate in parallel. By exploiting the speed of a hardware implementation to time-multiplex neurons, we will be able to test our approach on much larger networks of neurons; on the other hand, we are also looking to investigate the feasibility of evolving the genetic code in a real time hardware environment. The ultimate goal of this is to design a hardware implementation of CGPDN and the evolutionary system.

This implementation will allow us to realize large networks of neurons and analyse features of our model (e.g., the evolution of learning capabilities and the synergy between evolution and neuronal plasticity) that cannot be studied in small networks.

In addition to these features, CGPDN finds numerous applications in medicine that is pattern recognition for diagnosis of diseases, image processing, and in a variety of intelligent control systems.

Bibliography

Abdallah El Ali, L. B., Groen, I., Hermanides, E., Kool, W., Neville, D., & Rattner, K. (2008). *Forget-me-net: Overcoming catastrophic forgetting in backpropagation neural networks*. CSCA Summerschool.

Alberts, B., Johnson, A., Walter, P., Lewis, J., Raff, M., & Roberts, K. (2002). *Molecular biology of the cell* (3rd edn.). Garland Science.

Albesano, D., Gemello, R., Laface, P., Mana, F., & Scanzio, S. (2006). Adaptation of artificial neural networks avoiding catastrophic forgetting. *IJCNN* (pp. 1554–1561). IEEE.

Angeline, P. J., Saunders, G. M., & Pollack, J. B. (1993). An evolutionary algorithm that constructs recurrent neural networks. *IEEE Transactions on Neural Networks*, *5*, 54–65.

Arbab, M. A., Khan, G. M., & Sahibzada, A. M. (2014). Cardiac arrhythmia classification using cartesian genetic programming evolved artificial.

Back, T., Hoffmeister, F., & Schwefel, H. (1991). A survey of evolution strategies. In *Proceedings of the 4th International Conference on Genetic Algorithms* (Vol. 1802, pp. 2–9). Morgan Kaufmann.

Barricelli, N. A. (1954). Esempi numerici di processi di evoluzione. *Methodos,* 45–68.

Bentley, P. (2002). *Digital biology*. Simon and Schuster.

Bernstein, J. (1902). Untersuchungen zur thermodynamik der bioelektrischen strme. *Pfulger's Archives Gesellschaft Physiology*, *92*, 521–562.

Boers, E. J. W., & Kuiper, H. (1992). *Biological metaphors and the design of modular neural networks*. Masters thesis, Department of Computer Science and Department of Experimental and Theoretical Psychology, Leiden University.

Bongard, J. C. and Pfeifer, R. (2001). Repeated structure and dissociation of genotypic and phenotypic complexity in artificial ontogeny. In Spector et al. 2001 (pp. 829–836).

Braun, H., & Weisbrod, J. (1993). Evolving feedforward neural networks. In *Proceedings of ICANNGA93, International Conference on Artificial Neural Networks and Genetic Algorithms.* Innsbruck: Springer.

Cahill, A. (2010). *Catastrophic forgetting in reinforcement-learning environments*. A thesis submitted for the degree of Master of Science at the University of Otago, Dunedin, New Zealand.

Cajal, S. R., & y., (1894). The croonian reading: The fine structure of the centres nervous. *Proceedings of the Royal Society of London*, *55*, 444–468.

Cangelosi, A., Nolfi, S., & Parisi, D. (1994). Cell division and migration in a 'genotype' for neural networks. *Network-Computation in Neural Systems*, *5*, 497–515.

Carpenter, G. A., & Grossberg, S. (1988). The art of adaptive pattern recognition by a self-organizing neural network. *Computer*, *21*(3), 77–88.

Chalup, S. K. (2001). Issues of neurodevelopment in biological and artificial neural networks. In *Proceedings of the Fifth Biannual Conference on Artificial Neural Networks and Expert Systems (ANNES'2001),* 40–45.

G.M. Khan, *Evolution of Artificial Neural Development*, Studies in Computational Intelligence 725, https://doi.org/10.1007/978-3-319-67466-7

Chellapilla, K., & Fogel, D. B. (2001). Evolving an expert checkers playing program without using human expertise. *IEEE Transaction on Evolutionary Computation, 5*, 422–428.

Chen, X., & Hurst, S. (1982). A comparison of universal-logic-module realizations and their application in the synthesis of combinatorial and sequential logic networks. *IEEE Transactions on Computers, 31*, 140–147.

Cliff, D., & Miller, G. F. (1996). Co-evolution of pursuit and evasion ii: Simulation methods and results. In *Proceedings of the Fourth International Conference on Simulation of Adaptive Behavior* (pp. 506–515). MIT Press Bradford Books.

Cramer, N. L. (1985). A representation for the adaptive generation of simple sequential programs. In J. J. Grefenstette (Ed.), *Proceedings of an International Conference on Genetic Algorithms and the Applications*. Carnegie Mellon University.

Cunningham, P., Carney, J., & Jacob, S. (2000). Stability problems with artificial neural networks and the ensemble solution. *Artificial Intelligence in Medicine, 20*(3), 217–225.

Dalaert, F., & Beer, R. (1994). Towards an evolvable model of development for autonomous agent synthesis. In R. Brooks & P. Maes(Eds.), *Proceedings of the Fourth Conference on Artificial Life*. MIT Press.

Dasgupta, D., & McGregor, D. (1992). Designing application-specific neural networks using the structured genetic algorithm. *Proceedings of the International Conference on Combinations of Genetic Algorithms and Neural Networks,* 87–96.

DasGupta, B., & Schnitger, G. (1992). The power of approximating: A comparison of activation functions. *Advances in Neural Information Processing Systems, 5*, 363–374.

Dawkins, R., & Krebs, J. R. (1979). Arms races between and within species. *Proceedings of the Royal Society of London Series B, 205*, 489–511.

Debanne, D., Daoudal, G., Sourdet, V., & Russier, M. (2003). Brain plasticity and ion channels. *Journal of Physiology-Paris, 97*(4–6), 403–414.

Dimand, R. W., & Dimand, M. A. (1996). *A history of game theory: From the beginnings to 1945* (p. 1). Urbana: Routledge.

Dorffner, G. (1996). Neural networks for time series processing. *Neural Network World, 6*(4), 447–468.

Dorffner, G., & Porenta, G. (1994). On using feedforward neural networks for clinical diagnostic tasks. *Artificial Intelligence in Medicine, 6*(5), 417–435.

Downing, K. L. (2007). Supplementing evolutionary developmental systems with abstract models of neurogenesis. In *GECCO '07: Proceedings of the 9th Annual Conference on Genetic and Evolutionary Computation* (pp. 990–996). New York, NY, USA. ACM.

Elliot, W., & Elliot, D. (2001). *Biochemistry and molecular biology*. Oxford University Press.

Farkas, I., & Miikkulainen, R. (1999). Modeling the self-organization of directional selectivity in the primary visual cortex. In *International Conference on Artificial Neural Networks (ICANN '99)*, Edinburgh.

Federici, D. (2005). Evolving developing spiking neural networks. In *Proceedings of CEC 2005 IEEE Congress on Evolutionary Computation* (pp. 543–550).

Ferster, D., & Spruston, N. (1995). Cracking the neuronal code. *Science, 270*, 756–757.

Ficici, S. G., & Pollack, J. B. (1998). Challenges in co-evolutionary learning: Arms-race dynamics, open-endedness, and mediocre stable states. In C. Adami, R. Belew, H. Kitano, & C. Taylor (Eds.), *Artificial Life VI* (pp. 238–247). Cambridge MA: MIT Press.

Ficici, S. G., & Pollack, J. B. (2001). Pareto optimality in coevolutionary learning. In J. Kelemen (Ed.), *Sixth European conference on artificial life*. Berlin, New York: Springer.

Floreano, D., & Nolfi, S. (1997). God save the red queen! competition in co-evolutionary robotics. *Evolutionary Computation, 5*.

Fogel, D. (1998). *Evolutionary Computation: The Fossil Record*. Wiley-IEEE Press.

Fogel, L., Owens, A., & Walsh, M. (1966). *Artificial intelligence through simulated evolution*. Wiley.

Fogel, D. (2002). *Blondie24: Playing at the Edge of AI*. London, UK: Academic Press.

Forsyth, R. (1981). Beagle a darwinian approach to pattern recognition. *Kybernetes, 10*, 159–166.

French, R. (1991). Using semi-distributed representations to overcome catastrophic forgetting in connectionist networks. In *Proceedings of the Ninth Annual Conference of the Cognitive Science Society* (pp. 173–178). Hillsdale, NJ: LEA.
French, R. M. (1994). Catastrophic forgetting in connectionist networks: Causes, consequences and solutions. *Trends in Cognitive Sciences,* 128–135.
French, R. M. (1999). Catastrophic forgetting in connectionist networks: Causes, consequences and solutions. *Trends in Cognitive Sciences*, *3*(4), 128–135.
Frey, U., & Morris, R. (1997). Synaptic tagging and long-term potentiation. *Nature, 6, 385*(6616), 533–536.
Gaiarsa, J., Caillard, O., & Ben-Ari, Y. (2002). Long-term plasticity at gabaergic and glycinergic synapses: Mechanisms and functional significance. *Trends in Neurosciences*, *25*(11), 564–570.
Gerstner, W., & Kistler, W. (2002). *Spiking neuron models*. Cambridge University Press.
Gerstner, W., Kempter, R., Hemmen, L., Wagner, J., & Hebbian, H. (1999). Learning of pulse timing in the barn owl auditory system in maass. *Pulsed neural networks*.
Goldman, D. (1943). Potential, impedance and rectification in membranes. *Journal of General Physiology*, *27*, 37–60.
GoodMan, C., & Shatz, C. (1993). Developmental mechanisims that generated precise patterns of neuronal connectivity. *Cell*, *72*, 77–98.
Gopnic, A., Meltzoff, A., & Kuhl, P. (1999). *The scientist in the crib: What early learning tells us about the mind*. New York, NY: HarperCollins Publishers.
Graham, B. (2002). Multiple forms of activity-dependent plasticity enhance information transfer at a dynamic synapse. In J. R. Dorronsoro (Ed.), *ICANN 2002*. Berlin, Heidelberg: Springer (ICANN). *LNCS, 2415*, 45–50.
Greenough, W. T., Hwang, H. M., & Gorman, C. (1985). Evidence for active synapse formation or altered postsynaptic metabolism in visual cortex of rats reared in complex environments. *Proceeding of National Academy of Science United State of America*, *82*(13), 4549–4552.
Grossberg, S. (1980). How does a brain build a cognitive code? *Psychological Review*, *81*, 1–51.
Grossberg, S. (1982). *Studies of mind and Brain*. Boston, MA: D. Reidel Publishing Company.
Gruau, F., Whitley, D., & Pyeatt, L. (1996). A comparison between cellular encoding and direct encoding for genetic neural networks. In J. R. Koza, D. E. Goldberg, D. B. Fogel, & R. L. Riolo (Eds.), *Proceedings of the First Annual Conference* (pp. 81–89). MIT Press. (Genetic programming 1996).
Gruau, F. (1994). Automatic definition of modular neural networks. *Adaptive Behaviour*, *3*, 151–183.
Gurney, K. (1997). *An introduction to neural networks*. London: Routledge.
Hawkins, J. (2004). *On intelligence*. Times Books.
Haykin, S. (1998). *Neural Networks: A comprehensive foundation* (2nd Edn.). Prentice Hall.
Hebb, D. (1949). *The organization of behavior*. New York: Wiley.
Hertz, J., Krogh, A., & Palmer, R. (1991). *Introduction to the theory of neural computation*. Addison Wesley: Perseus Books.
Hillis, W. (1990). Co-evolving parasites improve simulated evolution as an optimization procedure. *Physica D: Nonlinear Phenomena*, *42*, 228–234.
Hillis, W. (1991). Co-evolving parasites improve simulated evolution as an optimization procedure. *Artificial life*, *2*, 313–324.
Hinton, G., & Plaut, D. (1987). Using fast weights to deblur old memories. In *Proceedings of the Ninth Annual Conference of the Cognitive Science Society* (pp. 177–186). Hillsdale, NJ: Erlbaum.
Hinton, G. E., Osindero, S., Welling, M., & Teh, Y. W. (2006). Unsupervised discovery of nonlinear structure using contrastive backpropagation. *Cognitive Science: A Multidisciplinary Journal*, *30*(4), 725–731.
Hodgkin, A. L., & Huxley, A. F. (1952). A quantitative description of membrane current and its application to conduction and excitation in nerve. *Journal of Physiology*, *463*, 391–407.
Holland, J. (1975b). *Adaptation in natural and artificial systems*. University of Michigan Press.
Holland, J. (1998). *Emergence: From chaos to order*. Oxford University Press.

Holland, J. (1975a). *Adaptation in natural and artificial system.* Ann Arbor: The University of Michigan Press.

Hopfield, J. J., & Tank, D. W. (1985). Neural computation of decisions in optimization problems. *Biological Cybernetics, 55*, 141–146.

Hopfield, J. (1982). Neural networks and physical systems with emergent collective computational abilities. *Proceedings of the National Academy of Sciences of the USA, 79*, 2554–2558.

Hornby, G. S., & Pollack, J. B. (2002). Creating High-level components with a generative representation for body-brain evolution. *Artificial Life, 8.*

Husbands, P., I., H., Cliff, D., & Miller, G., (1994). The use of genetic algorithms for the development of sensorimotor control systems. In P. Gaussier & J. D. Nicoud (Eds.), *From perception to action.* IEEE Press.

Husek, D., Frolov, A., Rezankova, H., & Snasel, V. (2002). Application of hopfieldlike neural networks to nonlinear factorization.

Huxley, A., & Stmpfli, R. (1949). Evidence for saltatory conduction in peripheral myelinated nervefibers. *Journal of Physiology, 108*, 315–39.

Irving, G., D. J., & Uiterwijk, J. (2000). Solving kalah. *International Computer Games Association (ICGA) Journal, 23*(3), 139–147.

Jakobi, N. (1995). *Harnessing morphogenesis, cognitive science research paper 423.* COGS: University of Sussex.

Jong, K. A. D. & Potter, M. A. (1995). Evolving complex structures via cooperative coevolution. In *Proceedings of the Fourth Annual Conference on Evolutionary Programming* (pp. 307–317). MIT Press.

Jordan, M. I. (1986). Attractor dynamics and parallellism in a connectionist sequential machine.

Juille, H., & Pollack, J. B. (1998). Coevolving the *ideal* trainer: Application to the discovery of cellular automata rules. In *Proceedings of the Third Annual Genetic Programming Conference.* Third Annual Genetic Programming Conference , Madison, Wisconsin.

Kandel, E. R., Schwartz, J. H., & Jessell, T. (2000). *Principles of neural science* (4th Edn.). McGraw-Hill.

Kaski, S., Kangas, J., & Kohonen, T. (1998). Bibliography of self-organizing map (som) papers: 1981–1997. *Neural Computing Surveys, 1*, 102–350.

Kelvin, W. (1855). On the theory of the electric telegraph. *Proceedings of the Royal Society, 7*, 382–99.

Kendall, G., & Whitwell, G. (2001). An evolutionary approach for the tuning of a chess evaluation function using population dynamics. *IEEE Congress on Evolutionary Computation (CEC 2001)*, 995–1002.

Khan, G. M., Ali, J., & Mahmud, S. A. (2014). Wind power forecasting? an application of machine learning in renewable energy. In *2014 International Joint Conference on Neural Networks (IJCNN)* (pp. 1130–1137). IEEE.

Khan, G. M., Khattak, A. R., Zafari, F., & Mahmud, S. A. (2013a). Electrical load forecasting using fast learning recurrent neural networks. In *The 2013 International Joint Conference on Neural Networks (IJCNN)* (pp. 1–6). IEEE.

Khan, G. M., Nayab, D., Mahmud, S. A., & Zafar, H. (2013b). Evolving dynamic forecasting model for foreign currency exchange rates using plastic neural networks. In *12th International Conference on Machine Learning and Applications (ICMLA) 2013* (Vol. 2, pp. 15–20). IEEE.

Khan, G. M., Ullah, F., & Mahmud, S. A. (2013c). Mpeg-4 internet traffic estimation using recurrent cgpann. In *International Conference on Engineering Applications of Neural Networks* (pp. 22–31). Springer.

Khan, M. M., Ahmad, A. M., Khan, G. M., & Miller, J. F. (2013d). Fast learning neural networks using cartesian genetic programming. *Neurocomputing, 121*, 274–289.

Khan, G., Miller, J., & Halliday, D. (2007). Coevolution of intelligent agents using cartesian genetic programming. *Proceedings of the 9th Annual Conference on Genetic and Evolutionary Computation,* 269–276.

Khan, G. M., & Zafari, F. (2016). Dynamic feedback neuro-evolutionary networks for forecasting the highly fluctuating electrical loads. *Genetic Programming and Evolvable Machines, 17*(4), 391–408.

Kirkpatric, J., Pascanu, R., Rabinowitz, N., Veness, J., Desjardins, G., Rusu, A. A., et al. (2017, March 28). Overcoming catastrophic forgetting in neural networks. *Proceedings of the National Academy of Sciences,* 3521–3526.

Kleim, J., Napper, R., Swain, R., Armstrong, K., Jones, T., & Greenough, W. (1998). Selective synaptic plasticity in the cerebellar cortex of the rat following complex motor learning. *Neurobiology of Learning and Memory, 69*, 274–289.

Koch, C., & Segev, I. (2000). The role of single neurons in information processing. *Nature Neuroscience Supplement, 3*, 1171–1177.

Kohonen, T. (1982). Self-organized formation of topologically correct feature maps. *Biological Cybernetics, 43*, 59–69.

Kohonen, T. (2001). *Self-organizing maps*. Berlin: Springer.

Kohonen, T., & Somervuo, P. (2002). How to make large self-organizing maps for nonvectorial data. *Neural Networks, 15*(8–9), 945–952.

Koza, J. R. (1990). *Genetic programming: A paradigm for genetically breeding populations of computer programs to solve problems*. Stanford University Computer Science Department technical report STAN-CS-90-1314. A thorough report, possibly used as a draft to his 1992 book.

Koza, J. (1992). *Genetic programming: On the programming of computers by means of natural selection*. MIT Press.

Koza, J. R. (1994). *Genetic programming II: Automatic discovery of reusable subprograms*. MIT Press.

Koza, J., Bennett, F., Andre, D., & Keane, M. (1999). *Genetic programming III: Darwinian invention and problem solving*. Morgan Kaufmann.

Koza, J., Keane, M., Streeter, M., Mydlowec, W., Yu, J., & Lanza, G. (2003). *Genetic programming IV: Routine human-competitive machine intelligence*. Kluwer Academic Publishers.

Krishnan, R. & Ciesielski, V. B. (1994). 2delta-gann: A new approach to training neural networks using genetic algorithms. *Proceedings of the Australian Conference on Neural Networks,* 194–197.

Kuffler, S. W., Nichols, J. G., & Martin, A. R. (1984). *From neuron to brain, a cellular approach to the function of the nervous system* (Second Edn.). Sinauer Press.

Kumar, S., & Bentley, B. J. (2003). *On growth, form and computers*. Academic Press.

Lee, C.-H., & Kim, J.-H. (1996). Evolutionary ordered neural network with a linked-list encoding scheme. *Proceedings of the 1996 IEEE International Conference on Evolutionary Computation,* 665–669.

Lieb, W., & Stein, W. (1986). *Chapter 2. Simple diffusion across the membrane barrier: Transport and diffusion across cell membranes*. San Diego: Academic Press.

Lillie, R. (1925). Factors affecting transmission and recovery in passive iron nerve model. *Journal of General Physiology, 7*, 473–507.

Lindenmeyer, A. (1968). Mathematical models for cellular interaction in development, parts i and ii. *Journal of Theoretical Biology, 18*, 280–315.

Lindgren, K., & Johansson, J. (2001). Coevolution of strategies in n-person prisoners dilemma. In J. Crutchfield & P. Schuster (Eds.), *Evolutionary dynamics—Exploring the interplay of selection, neutrality, accident, and function*. Reading, MA: Addison Wesley.

Lodish, H., Berk, A., Matsudaira, P., Kaiser, C., Krieger, M., Scott, M., et al. (2003). *Molecular cell biology*. W.H: Freeman.

London, M., & Husser, M. (2005). Dendritic computation. *Annual Review of Neuroscience, 28*, 503–532.

Lubberts, A., & Miikkulainen, R. (2001). Co-evolving a go-playing neural network. In R. Belew & H. Juille (Eds.), *Coevolution: Turning adaptive algorithms upon themselves* (pp. 14–19).

Maass, W., Schnitger, G., & Sontag, E. (1991). On the computational power of sigmoid versus boolean threshold circuits. *Proceedings of the 32nd Annual IEEE Symposium on Foundations of Computer Science,* 767–776.

Malenka, R., & Bear, M. (2004). Ltp and ltd: An embarrassment of riches. *Neuron, 44*(1), 5–21.

Mandischer, M. (1993). Representation and evolution of neural networks. In R. F. Albrecht, C. R. Reeves, & U. C. Steele(Eds.), *Artificial neural nets and genetic algorithms* (pp. 643–649).

Maniezzo, V. (1994). Genetic evolution of the topology and weight distribution of neural networks. *IEEE Transactions on Neural Networks, 5*(1), 39–53.

Marcus, G. F. (2001). Plasticity and nativism: Towards a resolution of an apparent paradox (pp. 368–382).

Martinetz, T., Berkovich, S., & Schulten, K. (1993). "neural gas" for vector quantization and its application to time-series prediction. *IEEE Transactions on Neural Networks, 4*, 558–569.

McCloskey, M., & Cohen, N. (1989). Catastrophic interference in connectionist networks: The sequential learning problem. *The Psychology of Learning and Motivation, 24*, 109–165.

McCulloch, & Pitts, W. (1943). A logical calculus of the ideas immanent in nervous activity. *The Bulletin of Mathematical Biophysics, 5*, 115–133.

Mehrtash, N., Jung, D., Hellmich, H., Schoenauer, T., Lu, V., & Klar, H. (2003). Synaptic plasticity in spiking neural networks (sp/sup 2/inn): A system approach. *IEEE Transaction on Neural Networks, 14*(5), 980–992.

Mel, B. (1994). Information processing in dendritic trees. *Neural Computation, 6*, 1031–1085.

Mermillod, M., Bugaiska, A., & BONIN, P. (2013). The stability-plasticity dilemma: Investigating the continuum from catastrophic forgetting to age-limited learning effects. *Frontiers in Psychology, 4*, 504.

Michael, S., Richard, B., & George, R. (1998). *Cognitive NeuroScience*. The Biology of the Mind: W.W.Norton & Company.

Miller, J. F. (1999). An empirical study of the efficiency of learning boolean functions using a cartesian genetic programming approach. In W. Banzhaf, J. Daida, A. E. Eiben, M. H. Garzon, V. Honavar, M. Jakiela, & R. E. Smith (Eds.), *Proceedings of the Genetic and Evolutionary Computation Conference* (Vol. 2, pp. 1135–1142). Orlando, Florida, USA. Morgan Kaufmann.

Miller, J. F. (2003). Evolving developmental programs for adaptation, morphogenesis and self-repair. In *Proceedings of the 7th European Conference on Advances in Artificial Life. LNAI* (Vol. 2801, pp. 289–298).

Miller, J. F. (2004). Evolving a self-repairing, self-regulating, french flag organism. In K. Deb et al. (Eds.), *Proceedings of GECCO* (Vol. 3102, pp. 129–139).

Miller, J. F., & Thomson, P. (2000). Cartesian genetic programming. In *Proceedings of the 3rd European Conference on Genetic Programming* (Vol. 1802, pp. 121–132).

Miller, J. F., Thomson, P., & Fogarty, T. C. (1997). Designing electronic circuits using evolutionary algorithms. arithmetic circuits: A case study. *Genetic algorithms and evolution strategies in engineering and computer science* (pp. 105–131). Wiley.

Miller, J. F., Vassilev, V. K., & Job, D. (2000). Principles in the evolutionary design of digital circuits-part i. *Genetic Programming, 1*(1/2), 7–35.

Moriarty, D., & Miikulainen, R. (1995). Discovering complex othello strategies through evolutionary neural networks. *Connection Science, 7*(3–4), 195–209.

Mummert, H., & Gradmann, D. (1991). Action potentials in acetabularia: measurement and simulation of voltage-gated fluxes. *Journal of Membrane Biology, 124*, 265–73.

Murray, S. (1993). *Neural networks for statistical modeling*. Van Nostrand Reinhold.

Nayab, D., Khan, G. M., & Mahmud, S. A. (2013). Prediction of foreign currency exchange rates using cgpann. In *International Conference on Engineering Applications of Neural Networks* (pp. 91–101). Springer.

Neumann, J., & V. (1928). Zur theorie der gesellschaftsspiele. *Mathematische Annalen, 100*, 295–320.

Nimchinsky, E., Sabatini, B., & Svoboda, K. (2002). Structure and function of dendritic spines. *Annual Review of Physiology, 64*, 313–53.

Noble, J. & Watson, R. A. (2001). Pareto coevolution: Using performance against coevolved opponents in a game as dimensions for parerto selection. In In et al., L. S. (Ed.), *Proceedings of the Genetic and Evolutionary Computation Conference (GECCO-2001)*. San Francisco, CA: Morgan Kaufmann.

Nolfi, S., & Parisi, D. (1995). Genotype for neural networks. In M. A. Arbib (Ed.), *Handbook of Brain theory and Neural Networks*. MIT Press.

Nolfi, S., Miglino, O., & Parisi, D. (1994). Phenotypic plasticity in evolving neural networks. In D. P. Gaussier & J. D. Nnicoud (Eds), *Proceedings of the international conference from perception to action*. IEEE Press.

Nolfi, S., & Floreano, D. (1998). Co-evolving predator and prey robots: Do 'arm races' arise in artificial evolution? *Artificial Life*, *4*, 311–335.

Ole-Marius, Moe-Helgesen, & Stranden, H. (2005). *Catastophic forgetting in neural networks*. NTNU.

Ooyen, V. A., & van Pelt, J. (1994). Activity-dependent outgrowth of neurons and overshoot phenomena in developing neural networks. *Journal of Theoretical Biology*, *167*, 27–43.

Opitz, D. W., & Shavlik, J. W. (1997). Connectionist theory refinement: Genetically searching the space of network topologies. *Journal of Artificial Intelligence Research*, *6*, 177–209.

Panchev, C., Wermter, S., & Chen, H. (2002). Spike-timing dependent competitive learning of integrate-and-fire neurons with active dendrites. In J. R. Dorronsoro (Ed.), *ICANN 2002*. Berlin Heidelberg: Springer (ICANN); (*LNCS, 2415*, 896–901).

Papadrakakis, M., Papadopoulos, V., & Lagaros, N. D. (1996). Structural reliability analyis of elastic-plastic structures using neural networks and monte carlo simulation. *Computer Methods in Applied Mechanics and Engineering*, *136*(1–2), 145–163.

Paredis, J. (1994a). Coevolutionary constraint satisfaction. In *Proceedings of the Third International Conference on Parallel Problem Solving from Nature* (Vol. 866, pp. 46–55). Springer.

Paredis, J. (1994b). Steps towards co-evolutionary classification neural networks. *Artificial Life iv*, *2*, 102–108.

Paredis, J. (1995). Coevolutionary computation. *Artificial Life*, *2*, 355–375.

Parisi, D., & Nolfi, S. (2001). Development in neural networks. In M. Patel, V. Honovar, & K. Balakrishnan (Eds.), *Advances in the evolutionary synthesis of intelligent agents*. MIT Press.

Parisi, D. (1997). Artificial life and higher level cognition. *Brain and Cognition*, *34*, 160–184.

Pollack, J., Blair, A., & Land, M. (1996). Coevolution of a backgammon player. In C. Langton (Ed.), *Proceedings artificial life 5*. MIT Press.

Pujol, J. C. F., & Poli, R. (1997). Evolution of the topology and the weights of neural networks using genetic programming with a dual representation. *Technical Report CSRP-97-7, School of Computer Science, The University of Birmingham, Birmingham B15 2TT, UK*.

Quartz, S., & Sejnowski, T. (1997). The neural basis of cognitive development: A constructivist manifesto. *Behavioral and Brain Sciences*, *20*, 537–556.

Rall, W. (1989). Cable theory for dendritic neurons. C. Koch & I. Segev (Eds.). *Methods in neuronal modeling: From synapses to networks* (p. 962).

Ratcliff, R. (1990). Connectionist models of recognition and memory: Constraints imposed by learning and forgetting functions. *Psychological Review*, *97*, 205–308.

Rechenberg, I. (1971). Evolutionsstrategie - optimierung technischer systeme nach prinzipien der biologischen evolution (phd thesis). *Reprinted by Fromman-Holzboog*.

Rechenberg, I. (1994). *Evolutionsstrategie '94*. Stuttgart: Frommann-Holzboog.

Rehman, M., Ali, J., Khan, G. M., & Mahmud, S. A. (2014a). Extracting trends ensembles in solar irradiance for green energy generation using neuro-evolution. In *IFIP International Conference on Artificial Intelligence Applications and Innovations* (pp. 456–465). Springer.

Rehman, M., Khan, G. M., & Mahmud, S. A. (2014b). Foreign currency exchange rates prediction using cgp and recurrent neural network. *IERI Procedia*, *10*, 239–244.

Richards, N., Moriarty, D., McQuesten, P., & Miikkulainen, R. (1998). Evolving neural networks to play go. In *7th International Conference on Genetic Algorithms*.

Ripley, B. D. (1996). *Pattern recognition and neural networks*. Cambridge University Press.

Roberts, P., & Bell, C. (2002). Spike-timing dependent synaptic plasticity in biological systems. *Biological Cybernetics*, *87*, 392–403.
Rose, S. (2003). *The making of memory: From molecules to mind.* Vintage.
Rosin, C. D. (1997). *Coevolutionary search among adversaries*. Ph.D. thesis, University of California, San Diego.
Rosin, C. D., & Belew, R. K. (1997). New methods for competitive evolution. *Evolutionary Computation, 5.*
Rumelhart, D. E., Hinton, G. E., & Williams, R. J. (1986). Learning representations by back-propagating errors. *Nature*, *323*, 533–536.
Russell, S., & Norvig, P. (1995). *Artificial Intelligence*. A Modern Approach: Prentice Hall.
Rust, A. G. & Adams, R. (1999). Developmental evolution of dendritic morphology in a multi-compartmental neuron model. In *Proceedings of the 9th International Conference on Artificial Neural Networks (ICANN'99)* (Vol. 1, pp. 383–388). IEEE.
Rust, A., Adams, R., & H., B. (2000). Evolutionary neural topiary: Growing and sculpting artificial neurons to order. In *Proceedings of the 7th International Conference on the Simulation and synthesis of Living Systems (ALife VII)* (pp. 146–150). MIT Press.
Sadeghi, B. (2000). A bp-neural network predictor model for plastic injection molding process. *Journal of Materials Processing Technology*, *103*(3), 411–416.
Samuel, A. (1959). Some studies in machine learning using the game of checkers. *IBM Journal of Research and Development*, *3*(3), 210–219.
Schaeffer, J., Herik, J., & V. D. (2002). *Chips challenging champions*. Amsterdam: Elsevier.
Schaeffer, J. (1996). *One jump ahead: Challenging human supremacy in checkers*. Berlin: Springer.
Schmidhuber, J. (1987). *Evolutionary principles in self-referential learning*. Diploma thesis, Institut f. Informatik, Tech. Univ. Munich.
Seipone, T., & Bullinaria, J. A. (2005). The evolution of minimal catastrophic forgetting in neural systems. In *Proceedings of the Twenty-Seventh Annual Conference of the Cognitive Science Society*. Mahwah, NJ: Lawrence Erlbaum Associates.
Shannon, C. (1950). Programming a computer for playing chess. *Philosophical Magazine*, *41*, 256–275.
Sharkey, N., & Sharkey, A. (1995). An analysis of catastrophic interference. *Connection Science, 7*(3–4), 301–330(30).
Shatz, J. C. (1994). Role for spontaneous neural activity in the patterning of connections between retina and lgn during visual system development. *Intenational Journal of Developmental Neuroscience*, *12*(6), 531–546.
Shepherd, G. (1990). *The synaptic organization of the brain*. Oxford Press.
Sims, K. (1994). Evolving 3d morphology and behavior by competition. In *Artificial life 4 proceedings* (pp. 28–39). MIT Press.
Sjoberg, J., Hjalmarsson, H., & Ljung, L. (1994). Neural networks in system identification.
Smith, S. (1980). *A Learning System Based on Genetic Adaptive Algorithms, PhD dissertation*. University of Pittsburgh.
Smythies, J. (2002). *The dynamic neuron*. BradFord.
Song, S., Miller, K., & Abbott, L. (2000). Competitive hebbian learning through spiketime - dependent synaptic plasticity.
Sordo, M. (2002). *Introduction to neural networks in healthcare*.
Spector, L. (1996). Simultaneous evolution of programs and their control structures. In P. J. Angeline & K. E. Kinnear, Jr. (Eds.), *Advances in genetic programming* (Vol. 2, pp. 137–154). Cambridge, MA, USA: MIT Press.
Spector, L., & Luke, S. (1996). Cultural transmission of information in genetic programming. In J. R. Koza, D. E. Goldberg, D. B. Fogel, & R. L. Riolo (Eds.), *Genetic programming 1996: Proceedings of the first annual conference* (pp. 209–214), Stanford University, CA, USA: MIT Press.
Stanley, K. O., & Miikkulainen, R. (2002). Evolving neural network through augmenting topologies. *Evolutionary Computation*, *10*(2), 99–127.

Stanley, K. O., & Miikkulainen, R. (2003). A taxonomy for artificial embryogeny. *Artificial Life, 9*(2), 93–130.
Stanley, K., & Miikkulainen, R. (2004). Competitive coevolution through evolutionary complexification. *Journal of Artificial Intelligence Research, 3*(21), 63–100.
Steinbach, H., & Spiegelman, S. (1943). The sodium and potassium balance in squid nerve axoplasm. *Journal of Cellular and Comparative Physiology, 22*, 187–96.
Stuart, G., Spruston, N., & Hausser, M. E. (2001). *Iterative broadening: Dendrites*. Oxford University Press.
Tasaki, I., & Takeuchi, T. (1942). Weitere studien ber den aktionsstrom der markhaltigen nervenfaser und ber die elektrosaltatorische bertragung des nervenimpulses. *Pfulger's Archives Gesellschaft Physiology, 245*, 764–82.
Tasaki, I. (1939). Electro-saltatory transmission of nerve impulse and effect of narcosis upon nerve fiber. *American Journal of Physiology, 127*, 211–27.
Terje, L. (2003). The discovery of long-term potentiation. *Philosophical Transactions of the Royal Society of London. Series B, Biological Sciences, 358*(1432), 617–20.
Thorpe, S., Delorme, A., & Van Rullen, R. (2001). Spike based strategies for rapid processing. *Neural Networks, 14*, 715–726.
Timothy, M. (1994). *Signal and image processing with neural networks*. Wiley.
Traub, R. (1977). Motoneurons of different geometry and the size principal. *Biological Cybernetics, 25*, 163–176.
Van Ooyen, A., & Pelt, J. (1994). Activity-dependent outgrowth of neurons and overshoot phenomena in developing neural networks. *Journal of Theoretical Biology, 167*, 27–43.
Van Rossum, M. C. W., Bi, G. Q., & Turrigiano, G. G. (2000). Stable hebbian learning from spike timing-dependent plasticity. *Journal of Neuroscience, 20*, 8812–8821.
Van Valin, L. (1973). A new evolutionary law. *Evolution Theory, 1*, 130.
Vassilev, V. K., & Miller, J. F. (2000). The advantages of landscape neutrality in digital circuit evolution. In *Proceedings of the 3rd ICES* (Vol. 1801, pp. 252–263). Springer.
Walker, J. A., & Miller, J. F. (2004). Evolution and acquisition of modules in cartesian genetic programming. In *Proceedings of the 7th EuroGP* (Vol. 3003, pp. 187–197). LNCS, Springer.
Walker, J., & Miller, J. (2008). The automatic acquisition, evolution and reuse of modules in cartesian genetic programming. *IEEE Transactions on Evolutionary Computation, 12*.
Wee-Chong, O. & Yew-Jin, L. (2003). An investigation on piece differential information in coevolution on games using kalah. In *Proceedings of the 2003 Congress on Evolutionary Computation CEC2003* (pp. 1632–1638). IEEE Press.
Whitley, D., & Hanson, T. (1989). Optimizing neural network using faster more accurated genetic search. In *Proceeding of Third International Conference on Genetic Algorithms* (pp. 391–396). Morgan Kaufman.
Wood, D. C. (1988). Habituation in stentor produced by mechanoreceptor channel modification. *Journal of Neuroscience, 2254*(8),
Yao, X. (1999). Evolving artificial neural networks. *Proceedings of the IEEE, 87*(9), 1423–1447.
Yao, X., & Liu, Y. (1996). Towards designing artificial neural networks by evolution. *Applied Mathematics and Computation, 91*(1), 83–90.
Yob, G. (1975). Hunt the wumpus. *Creative Computing,* 51–54.
Yu, T., & Miller, J. (2001). Neutrality and the evolvability of boolean function landscape. In *Proceedings of the 4th EuroGP,* (pp. 204–217). Springer.
Yu, T., & Miller, J. (2002). Finding needles in haystacks is not hard with neutrality. In *Proceedings of the 5th EuroGP* (Vol. 1801, pp. 13–25). Springer.
Zhang, B., & Muhlenbein, H. (1993). Evolving optimal neural networks using genetic algorithms with occams razor. *Complex Systems, 7*, 199–220.

Printed in the United States
By Bookmasters